John Wesley, William Hawes

An examination of the Rev. Mr. John Wesley's Primitive physic

Interspersed with medical remarks and practical observations

John Wesley, William Hawes

An examination of the Rev. Mr. John Wesley's Primitive physic
Interspersed with medical remarks and practical observations

ISBN/EAN: 9783337202378

Printed in Europe, USA, Canada, Australia, Japan

Cover: Foto ©berggeist007 / pixelio.de

More available books at **www.hansebooks.com**

A N

EXAMINATION

O F

The Rev. Mr. JOHN WESLEY's

PRIMITIVE PHYSIC.

[PRICE ONE SHILLING AND SIX-PENCE.]

A N

EXAMINATION

O F

The Rev. Mr. JOHN WESLEY's

PRIMITIVE PHYSIC:

SHEWING

That a great Number of the Prefcriptions therein contained, are
founded on Ignorance of the MEDICAL ART, and of the
Power and Operation of Medicines; and that it is a Publication calculated to do effential Injury to the Health of
thofe Perfons who may place Confidence in it.

INTERSPERSED WITH

MEDICAL REMARKS

A N D

PRACTICAL OBSERVATIONS.

By W. HAWES, APOTHECARY.

*Medico diligenti, priufquam conetur ægro adhibere medicinam, non
folum morbus ejus, cui mederi volet, fed etiam confuetudo valentis,
et natura corporis cognofcenda eft.*

Cicero de Oratore, lib. ii. cap. 44.

LONDON:
Printed for the AUTHOR; and fold by J. DODSLEY, Pall-
Mall; T. CADELL, Strand; B. JOHNSON, St. Paul's
Church-yard; and W. FOX, Holborn.
MDCCLXXVI.

PREFACE.

THE writer of the following pages was induced to communicate them to the world, from a defire to prevent the public from being longer impofed on, by an injudicious collection of pretended remedies for almoft every diforder that can affect the human frame, and which has been publifhed by Mr. John Wefley, under the title of *Primitive Phyfic*. This writer, or rather compiler, has laboured to give mankind the moft unfavourable ideas of the practitioners in phyfic and pharmacy. The *phyficians* he reprefents as engaged in a combination to render their art as myfte-

rious

rious as poffible, in order to make themfelves the more neceffary, and to increafe the gains of their profeffion. While *apothecaries*, he infinuates, make little fcruple of adminiftering drugs not contained in the prefcription of the phyfician, becaufe they are more cheap, or fuch as are ftale and perifhed, to the ruin of *many conftitutions*, and to the lofs of *many valuable lives*. And from this account it fhould feem, that phyficians and apothecaries not unfrequently combine together, for no other purpofe than to plunder the patient, and to encreafe or prolong his mifery and his difeafe. " Experience fhews, fays " he, that one thing will cure moft " diforders, at leaft as well as twenty " put together. Then why do you " add the other nineteen? Only to " fwell the apothecary's bill : nay poffibly to prolong the diftemper, that

" the

" the doctor and he may divide the
" fpoil *."

This reprefentation of the gentlemen of the faculty may poffibly not
be thought very candid, nor very
equitable : and if Mr. Wefley's character and conduct, as a divine, a po-
litician, and a practitioner in phyfic,
were to be examined with the fame
degree of candour that he hath exer-
cifed towards others, he would certainly not appear in the moft advantageous light. At leaft it would
be manifeft, that he was far enough
from *perfection*, though that is a doc-
trine for which he is well known to
be a very zealous advocate. But,
perhaps, thofe who are not thorough-
initiated in Mr. Wefley's peculiar te-
nets, may not have a proper idea of
what thofe qualities are which are neceffary to conftitute a perfect man.

* Pref. to *Primitive Phyfic*, p. xiv. and alfo p. xxvii.

　　　　　　　　　　　　　　It

It is certain, that if Mr. Wesley be of this character, a regard to truth is not neceffary to it : of which the Rev. Mr. Evans of Briftol can afford ample teftimony †.

But however uncandid, unfair, or unjuft, Mr. Wesley's reprefentation of the gentlemen of the faculty may be, it feemed neceffary to promote the fale of his *Primitive Phyfic.* And in this his views appear to have been anfwered ; *fixteen editions* at leaft having been printed of this compilation ;

† *Vide* the fecond edition of Mr. Evan's letter to Mr. John Wefley, in which he has been convicted of premeditated falfehood, upon the clearft and moft unexceptionable evidence. Mr. Wefley's attempt towards a defence upon this fubject in the news-papers, ferves, *if poffible*, to render him ftill more contemptible.

The writer of this meddles not with political difputes, but takes the liberty to obferve, that fome regard to truth was thought neceffary, in old fafhioned fyftems, to conftitute the character of an honeft man, of whatever party he might be.

and

and that this large fale has not arifen from the merit of the performance, will, I am confident, be acknowledged by every man fkilled in the treatment of difeafes.

The practice of phyfic, according to Mr. Wefley's ideas, is a very eafy art. For, he informs us, " neither " the knowledge of *aftrology, aftronomy,* " *natural philofophy,* nor even *anatomy* " itfelf, is abfolutely neceffary to the " *quick* and *effectual cure* of *moft difeafes* " incident to human bodies : nor yet " any chymical, or exotic, or com- " pounded medicine, but a fingle plant " or fruit duly applied. So that *every* " *man of common fenfe* (unlefs in fome " rare cafes) may prefcribe either to " himfelf or his neighbour ; and may " be *very fecure* from doing harm, even " where he can do no good *." But

* Pref. to *Primitive Phyfic,* p. xi.

the

the facility of curing difeafes was not, it feems, fufficiently underftood, till the appearance of Mr. Wefley's Primitive Phyfic. And when this performance was announced to the world, every man who purchafed it, had " a " phyfician always in his houfe, and " one that attends without fee or re- " ward."

Mr. Wefley's performance would, indeed, have been a very valuable acquifition to the public, if it could really have qualified every man of common fenfe " to prefcribe to his " family as well as himfelf." But the truth is, that thofe who rely on Mr. Wefley's pamphlet, will often be led to trifle with the moft dangerous difeafes, and while they are forming vain expectations of obtaining relief from his infignificant prefcriptions, may be led to neglect timely application for

real

real and effectual affiftance, and thereby fuffer irreparable mifchief.

Mr. Wefley's pretended remedies
are of various kinds; great numbers
have no mark of diftinction; but he
has befides thefe *tried* remedies, *infallible* remedies, and a third--fort,
which he prefers to all the reft (being
probably more than infallible) and
which are marked with an *afterifk*.
But if the public are led to form a
juft eftimate of the merit of Mr. Wefley's Primitive Phyfic, they will place
little confidence in any remedies which
have no better authority than his recommendation, whether they are
marked *tried*, *infallible*, or diftinguifhed by an *afterifk*.

It was not the intention of the
writer of this piece, to fit down merely with a view to oppofe Mr. Wefley,
or to cavil at his publication. But he
wifhed

wifhed to be of fervice to his fellow creatures. He has, therefore, inter-fperfed occafional remarks on feveral of the difeafes for which Mr. Wefley has attempted to prefcribe : and if any of his obfervations fhould prove beneficial to mankind, it will afford him great pleafure, as his higheft ambition is to be ufeful in his profef-fion.

A N

A N

EXAMINATION, &c.

THE recipes contained in Mr. Wesley's *Primitive Physic*, are one thousand and twelve; they are therefore too numerous to be particularly animadverted on: but, from the remarks which will here be made on many of them, it will, it is presumed, be sufficiently apparent, that no person can, with any degree of safety, rely on a compilation so extremely injudicious; the pretended remedies contained therein, being often of no use, and those which might be of utility, generally unattended with such directions, or regard to times and circumstances, as would be necessary to render them efficacious; and indeed, often calculated only to produce the most dangerous and fatal effects.

Those recipes, contained in Mr. Wesley's book, on which I shall make remarks, will be taken in the order in which they lie in his pamphlet, and referred to by the numbers or figures which he has affixed to them. When I give his recipes, or make quotations from him, which I shall frequently do, his words will be distinguished by the Italic character. And as my reasons for this publication have already been given in the preface, I shall, without further introduction or apology, proceed to an examination of that profundity of medical skill and knowledge, which are contained in this incomparable system of *Primitive Physic*.

B

Abortion,

Abortion to prevent.

No. 1. Use a decoction of lignum guiacum; boiling an ounce in a quart of water.

Of all the remedies to prevent abortion, this is one of the moſt improper; for if it were to produce any effect, it would be the very reverſe: as this wood contains a ſtimulating eſſential oil, it would be diſpoſed to increaſe the action of the heart and arterial ſyſtem; ſo that this medicine, in all probability, will make the woman more liable to abort.

In a ſanguine habit, let blood.

Mr. Weſley does not here give the leaſt information, by which the female might know whether bleeding would be a proper remedy; nor does he conſider, that not one in a hundred of his readers know the meaning of the word *ſanguine*. Indeed, he ſeems reſolved, at his firſt ſetting out, to give the public a ſatisfactory evidence of his total want of medical knowledge. Another man would have ſaid, if there is a full, ſtrong pulſe, then bleeding in ſmall quantities, and at ſtated intervals, according to the ſtrength and other circumſtances of the patient, will diminiſh the quantity of blood in the veſſels, and thus leſſen the force of the circulation in the uterus, and ſo prevent abortion.

For an Ague.

No. 3. Go into the cold bath juſt before the cold fit.

There are many caſes and circumſtances in which the *cold bath* would be abſolutely improper, as well as dangerous; and thoſe are by no means cleared up in another part of Mr. W.'s pamphlet, where he gives ſome general directions to thoſe who are to uſe it.

No.

No. 11. *Make six middling pills of cobwebs. Take one a little before the cold fit ; two a little before the next fit ; (suppose the next day) the other three, if need be, a little before the third fit.—This seldom fails.*

Here Mr. W. appears to have excelled himself; he orders his cobwebs to be made into pills, but he does not reflect that there muſt be ſome viſcid ſubſtance added, to form a dry, light matter into pills ; ſo that it is to be preſumed Mr. W. is at the trouble of making the cobwebs into pills himſelf. But as the mind has a wonderful effect on the body, and in no diſeaſe more than the preſent, I would recommend to Mr. W. to have his patient, *a little before the co d fit,* carried into a room where this wonder-working remedy hangs in cluſters from the cieling ; here the imagination would have its full force, and aſtoniſhing cures perhaps be performed.

No. 12. *Two tea ſpoonfuls of ſal prunella, an hour before the fit.*

Mr. W. pays no attention whatever to the doſe of a very powerful neutral ſalt ; ſome tea ſpoons may contain *one,* others, *two ſcruples ;* ſurely, more preciſion was neceſſary ; but, perhaps, like the verdigris, this may be an error of the preſs, which has gone through ſix, or peradventure ſixteen editions.

No. 17. *Eat a lemon, rind and all.*

This preſcription, which appears at firſt ſight very ſimple, is exceedingly compounded, as in this remedy is contained aceſcent, mucilaginous, bitter, and watery juices. Further, no attention whatever is paid to the ſize of the lemon which is to be *eat ;* ſome weigh four ounces, others only half an ounce, or ſix drams ; and yet Mr. W. adviſes a *lemon, rind and all,* to be taken for the cure of an ague.

No. 19. *In the hot fit, take* 10, 12, *or* 15 *drops of laudanum.*

B 2

This

This is a very dangerous remedy, and, I believe, never advised before, in the hot fit of an intermittent fever; as, in all probabily, it would heat the system much more, endanger delirium, and convert the present disease into a continued fever, which every good practitioner is anxious to avoid.

No. 21. *Boil a handful of rib wort in whey, drink this warm an hour before the fit comes, and lie down and sweat.*

Mr. W. has a prodigious command over the constitution; it obeys his nod, and is subservient to his will; producing sweat, and the removal of diseases, appear to be the easiest things in the world with him.

No. 22. *A tea spoonful of salt of tartar, in spring water. This also cures double tertians, triple quartans, long lasting fevers, and most diseases arising from obstructions, especially if sena be premised twice or thrice.*

There cannot be put together an assemblage of greater absurdities; as this medicine can have no other tendency than to cause the afflicted to trifle with acute diseases. In fact, in this one section is included the greatest part of the diseases incident to the human body, as the celebrated BOERHAVE says, that most diseases arise from obstruction. Mr. W.'s assertion, therefore, that this alkaline salt will cure so great a variety of disorders, must certainly be a matter of very serious consequence, as so great a number of complaints come under his description, in which salt of tartar could produce no good effect; and thus the greatest injury must often be received by those deluded persons, who are prevented, by their ill-placed confidence in this recipe of Mr. Wesley's from having recourse to such medicines as would effectually relieve them.

No. 23. *Before, yea, in the midſt of the fit, take twenty drops of ſpirit of ſulphur, in a pint of cold water.*

Before, yea, even in the midſt of the fit, Mr. W. is of opinion, that, *twenty drops of ſpirit of ſulphur,* ſhould be taken *in a pint of cold water.* This can be of no uſe whatever ; *yea,* it may be ſometimes injurious. But as no mark is affixed to this preſcription ; neither an afteriſk, nor the mark of infallibility ; Mr. W. may poſſibly be willing that this recipe ſhould not be numbered among thoſe of the higheſt excellence.

No. 25. *Apply to each wriſt a plaiſter of treacle and ſoot.—Tried.*

As the word *tried* is affixed to this footy application, it may be preſumed that Mr. W. or his chimney-ſweeper, have experienced its efficacy.

A Double Tertian.

No. 27. *Take, before the fit, (after a purge or two) three ounces of cichory water, half a drachm of ſalt of wormwood, and fifteen drops of ſpirit of ſulphur.*

A very inelegant and unpleaſant ſaline draught.

No. 28. *To perfect the cure, on the fourth day after you miſs the fit, take two drachms of ſena, half a drachm of ſalt of tartar, infuſed all night in four ounces of cichory water.*

If Mr. W. has any view in this preſcription, he intends it as a purging remedy ; but repeated experience has ſhewn purgatives to be very injurious after all kinds of agues, and the moſt probable method to produce a relapſe.

A Quar-

A Quartan Ague.

No. 29. *Apply to the future of the head, when the fit is coming, wall July flowers, beating together the leaves and flowers with a little salt.*

It muſt be a ſubject of lamentation, that this wonderful remedy can only be procured at a certain ſeaſon of the year ; but it may be ſome abatement of our grief to recollect, that the loſs of this pretty flower may be ſupplied by the tulip, pink, &c. and indeed any of them, applied to the noſtrils, will be productive of effects on the ſyſtem, equally aſtoniſhing.

No. 35. *For a tertian or quartan, vomit an hour after the cold fit begins.*

When Mr. W. by an extraordinary fatality, hits upon a good remedy, he generally takes care to prevent its being of real ſervice to the patient, by directing it to be adminiſtered injudiciouſly and improperly. At the attack of acute diſeaſes, the matters contained in the ſtomach, inſtead of going through the digeſtive proceſs, become often putrid, acid, &c. which increaſes the ſymptoms of the diſeaſe. Practitioners, therefore, at the onſet of fevers, and other diſeaſes, have found it uſeful to clear the ſtomach of offending ſubſtances, by an emetic; and if any other good effect is expected from the vomiting, it is generally adviſed a little before the attack, as it has ſometimes prevented the fit coming on ; but Mr. W. in contradiction to common experience, and common ſenſe, adviſes his *vomit to be given an hour after the cold fit begins.* Further, as the *Primitive Phyſic* is intended chiefly for the unlearned, and as there are many kinds of ſubſtances which occaſion vomiting, would it not have been prudent in Mr. W. to have mentioned which deſerved the preference, and what would be the ſuitable doſe, and the fluid proper to work off the vomit ?

No.

No. 36. *Drink every morning a gill of white wine, wherein half a sliced orange is boiled.*

In this generous prescription there is no direction whether the person should drink a gill of wine for a week, a month, or a year; but it may not be improper to observe, that if a weakly and delicate woman drinks four ounces of wine every morning for any length of time, that, when it is left off, her spirits will flag for want of the stimulus, and thus an excellent woman be unhappily converted into a dram or wine drinker.

No. 39. *Take ten grains of powdered saffron before the fit, in a glass of white wine.*

Recent experience has convinced me, that saffron may be exhibited in much larger doses then Mr. W. prescribes, without producing any medicinal effects.

St. Anthony's Fire.

No. 41. *Take a glass of tar water warm in bed, every hour, washing the part with the same.*

The disease for which Mr. W. is here pretending to prescribe a remedy, is the erisipetalous inflammation, or St. Anthony's fire ; and surely then, a hot stimulating substance, which would encrease the burning heat, and all the symptoms of this troublesome and painful disease, is not very proper to be applied.   Mr. W. has however thought proper to recommend it, but Dr. Lewis, who is a good physician, as well as a judicious writer, says, " That all the turpentines are " hot and stimulating, they are given where inflam- " matory symptoms do not forbid their use ; and that " TAR differs from the turpentines, or native resi- " nous juices of the trees, in consequence of having " received a disagreeable empyreumatic impression " from the fire."

No. 42. *Drink just so much sea-water as does not vomit or purge, every morning for seven days: this is*
the

the proper meafure in whatever cafe. It feldom fails.

At No. 41, the patient was to drink *tar-water*, a very heating medicine; immediately after, in the fame difeafe, *fea-water* is prefcribed, which in confequence of the neutral and earthy falts diffolved in it, is a very cooling remedy. Mr. W. whofe medical talents are not of the ordinary kind, undertakes to cure the very fame difeafe both by *hot* and *cold* remedies.

No. 45. *Take two or three gentle purges.—No fever bears repeated purges better than this, efpecially when it affects the body.*

As this is an inflammation peculiar to weak and irritable habits, thofe labouring under this difeafe are not well able to bear evacuation; and indeed there are few diforders where it is more neceffary to keep up the ftrength of the patient than the prefent.

Apoplexy.

No. 49. *To prevent, ufe the cold bath, and drink only water.*

As the apoplexy is a difeafe which carries off great numbers, fhould not any perfon who intended to give medical advice, have been more explicit? Mr. W. fays, in his note, that an *apoplexy is a total lofs of all fenfe and voluntary motion, commonly attended with a ftrong pulfe, hard breathing and fnorting.* It is evident, from this definition of the difeafe, the fanguineous apoplexy is intended. Now I will venture to affirm, that immerging fuddenly in the cold bath, will be one of the moft likely means of reproducing this dangerous and often fatal difeafe. Here is no attention paid to age, fex, conftitution, or other circumftances, though they are abfolutely neceffary to be attended to, in a complaint of fo ferious a nature.

No. 58.

No. 50. *In the fit put a handful of salt into a pint of cold water, and if possible pour it down the throat of the patient.*

Mr. Wesley here says, that *if possible* in the fit of apoplexy, a pint of salt and water should be poured down the throat of the patient ; *and then he will immediately come to himself.* But if he had not been totally ignorant of the disease, or if he had understood his own definition of it, (See No. 49.) he would have known that it was totally impossible to force down any quantity of fluid during the fit. He says himself, that the disease is attended with *a loss of all sense and voluntary motion.* Now, is not the action of deglutition a voluntary motion, and can it be restored any other way, than by removing the disease ?

No. 51. *Fill the mouth with salt.*

Mr.W. here recommends filling the mouth with salt ; but the most likely consequence of this would be, *killing the patient,* by the stoppage of all respiration.

No. 52. *Blow powder of white hellebore up the nose.*

One of the most stimulating errhines in the whole *materia medica,* is here ordered to be blown up the nostrils, in the sanguineous apoplexy ; than which, nothing can be more improper or dangerous.

Fix a cupping glass, without scarifying, to the nape of the neck, and another to each shoulder,

One of Mr. Wesley's shining qualities, is the adroitness with which he renders a good remedy inefficacious, whenever he happens to blunder upon one. Here, the cupping glasses recommended are very proper ; but the directing them to be applied without scarifying, is in the highest degree absurd. By the scarification, the blood vessels would have been unloaded of their contents, and the pressure upon the brain taken off ; and therefore, what the Author of the *Primitive Physic* advises to be avoid-

C   ed,

ed, would be the moſt probable method of reſtoring the patient.

No. 53. *If the fit be ſoon after a meal, do not bleed, but vomit.*

Theſe directions are eſſentially wrong, and if purſued, might be fatal to many perſons who would be recovered.

No. 54. *Rub the head, feet, and hands ſtrongly, and let two ſtrong men carry the patient upright, backward and forward about the room.*

This advice is vague and inſignificant, as it is ſuffering people to trifle in a diſeaſe which requires the moſt expeditious methods of relief.

No. 55. *A ſeton in the neck, with a low diet, has often prevented a relapſe.*

This advice for preventing a relapſe, is very proper ; but the patient muſt firſt be brought out of the apoplectic fit, which he never can be, by any of Mr. W's preſcriptions.

But ſend for a good phyſician immediately.

Theſe words of Mr. W. are contained in the latter part of the fifty-ſecond ſection; after preſcribing ſeveral of the moſt abſurd and prepoſterous remedies that could eaſily enter the mind of man, he adviſes a good phyſician to be ſent for. The writer hopes that this is the only part of Mr. W's advice, to which any regard will be paid, in ſo dangerous a diſeaſe ; where the omiſſion of the application of the proper and judicious remedies, *only* for a few minutes, may be the cauſe of the death of the patient,

The Aſthma.

No. 57. *Take a pint of cold water every night, as you lie down in bed.*

No. 58.

No. 58. *A pint of cold water every morning, wafh-ing the head therein immediately after, and ufing the cold bath once a fortnight.*

No. 60. *Half a pint of tar water twice a day.*

No. 61. *Drink fea water every morning.*

No. 62. *Live a fortnight on boiled carrots.—It feldom fails.*

All thefe pretended remedies for afthmatic diforders, are fo inadequate to the purpofe, and manifeftly fo inefficacious, that they *only* do mifchief, by preventing thofe who confide in them from applying for fuch advice as might afford them real relief.

No. 65. *For prefent relief, vomit with a quart or more of warm water. The more you drink of it, the better.*

In fits of the afthma, the lungs are often greatly loaded and diftended with blood ; fo that vomiting, by whatever means excited, may be productive of much mifchief. Mr. W. in his *Primitive Phyfic,* gives no rules refpecting times or circumftances ; but remedies are to be ufed *indifcriminately,* at all times, and in all circumftances. Here, however, I muft take the liberty of obferving, that I have found, from repeated experience in the various methods of reftoring health, that much depends on remedies being exhibited at the proper period of a difeafe ; and therefore they lofe much of their efficacy, when exhibited by a perfon who is inattentive to a proper diftinction of thefe particulars, which, though feemingly trifling, are really of the utmoft importance.

A dry, or convulfive Afthma.

No. 66. *Juice of radifhes relieves much.*

No. 67. *A cup of ftrong coffee.*

 No. 69.

No. 69. *A tea made with hyſſop, or ground-ivy, or daiſy flowers and liquorice.*

No. 70. *A pint of new milk, morning and evening.* —*This has cured an inveterate aſthma.*

No. 71. *Uſe the cold bath thrice a week.*

No. 72. *Beat fine ſaffron ſmall, and take eight or ten grains every night.*

A very dangerous diſeaſe, rendered more ſo, by ſeveral trifling and inſignificant preſcriptions.

No. 73. *Dry and powder a toad, make it into pills, and take one every hour, till the convulſions ceaſe.*

Of all Mr. W.'s remedies for the convulſive aſthma, *powder of toad* is the moſt curious; but it is ſuited to the credulity of the frequenters of the Foundery.

No. 74. *Take from three to five grains of ipecacuanha every morning ; or from five to ten grains every other evening. Do this, if need be, for a month or ſix weeks. Five grains uſually vomit. In a violent fit, take a ſcruple inſtantly.*

It is always adviſeable, before the exhibition of vomits, to examine the pulſe and the ſtate of the conſtitution in general; for, if the blood-veſſels ſhould be loaded, the action of vomiting may perhaps endanger a rupture, and inſtantly prove fatal.

Bleeding at the Noſe, (to prevent).

No. 77. *Drink whey largely every morning, and eat much of raiſins.*

No. 78. *To cure it, apply to the neck, behind, and on each ſide, a cloth dipped in cold water.*

No. 79. *Waſh the temples, noſe, and neck with vinegar.*

No. 80. *Snuff up vinegar and water.*

No. 81.

No. 81. *Chew nettle root, spitting out the juice.*

No. 82. *Put up the nostrils powdered betony, with a little salt.*

No. 83. *Hold a red hot poker under the nose.*

No. 84. *Steep a rag in sharp vinegar, burn it, and blow it up the nose with a quill.*

No. 85. *In a violent case, go into a pond or river.*

The *red-hot poker* prescription (No. 83.) is undoubtedly new; and I am confident no one will dispute the honour of its invention with Mr. Wesley. I shall, however, beg leave to recommend this caution in the use of it, that no one should attempt the application, who has not a very steady hand, lest the patient should bear the marks of his effectual cure, in a manner that might induce a wicked world to think, the case had been such as required the adhibition of Leake's pills, rather than Wesley's poker; nor could an accident of this kind easily be remedied; as, I believe, Mr. *Patence*, the only gentleman, who, in this age, professes the Taliacotian art, now no longer carries on his nose-making and nose-mending manufactory.

But, to be serious; an hæmorrhage from the nose, is, in general, a very salutary effort of nature, to empty the loaded vessels of the head; so that such discharges of blood should by no means be hastily suppressed, as very large quantities may be thus slowly evacuated, without inducing much weakness in the system; so that this accidental hæmorrhage tends greatly to relieve, and often to cure stubborn disorders of the head, eyes, &c. whereas, if imprudently checked by astringents, internally or externally applied, such stoppage of the flux of the blood, may often be productive of inflammation of the neighbouring parts, and sometimes even apoplexy and palsy may be the consequence of such injudicious prescriptions as are given in the *Primitive Physic*.

Spitting

Spitting of Blood.

No. 93. *Take half a pint of stewed prunes for two or three nights.*

No. 94. *A glass of decoction of onions.*

No. 95. *Two spoonfuls of juice of nettles every morning, and a large cup of decoction of nettles every night.*

No. 96. *Take frequently a spoonful of the juice of nettles and plantane.*

No. 97. *Three spoonfuls of sage juice in a little honey.*

No. 98. *Half a tea spoonful of Barbadoes tar on sugar at night.*

Instead of making a comment on each of these remedies, which the writer is thoroughly convinced can be of little or no use in a spitting of blood ; he will content himself with observing, that it requires more medical skill than Mr. W. seems possessed of, to discover whether the discharge of blood issues from the mouth, the lungs, or the stomach ; and then it is necessary to be so far acquainted with the circulation as to be capable of judging whether the discharge is arterial or venous blood ; and further, to understand so much of the history of diseases, as to know the cause of the hæmorrhage, that is, whether the bleeding arises from an increased action of the arteries, from a relaxation, or from a rupture of the vessels.

Vomiting Blood.

No. 99. *Take two spoonfuls of nettle juice.*

No. 100. *One spoonful of the juice of quinces.*

No. 101. *A quarter of a pint of decoction of nettles and plantane, two or three times a day.*

Hæmorr-

Hæmorrhages, from whatever part they arife, are difpofed to continue till they prove fatal, or, which is much more common, they naturally ceafe; for when the veffels are fufficiently emptied of their contents, they are of courfe difpofed to contract, and no more blood is thrown out. And the reader may be affured, that one or other of thefe terminations muft be the event, if any reliance is placed on Mr. W.'s internal or external remedies. for the greater part of them do not bid fair to be of any advantage in the different bleedings advifed for.

The obfervations I have made on the various hæmorrhages which Mr. W. treats of, fhould be well confidered in the treatment of them; for if aftringents and repellents are indifcreetly and injudicioufly prefcribed, they may do irreparable mifchief, in cafes where a mere difcharge of blood would have proved falutary.

Blifters.

No. 104. *On the feet, occafioned by walking, are cured by drawing a needle-full of worfted through them, clip it off at both ends, and leave it till the fkin peels off.*

In this cafe it would be better, that no wound fhould be made, as the watery fluids extravafated from the ftimulus of walking, will generally be abforbed during a night's reft, and the bliftered part reftored to its natural ftate.

Boils.

No. 105, *to* 108. *Are feveral external applications to promote fuppuration. 'Tis proper to purge alfo.*

If the habit of body fhould be in a good ftate when thefe external inflammations arife, the fuppuration will be good, and the boils heal readily; but on
the

the other hand, if the conftitution is fcorbutic, or the juices altered from their natural ftate, neither the above external remedies, nor purgatives, will compleat the cure.

Hard Breafts.

No. 109. *Apply turnips roafted till foft, then mafhed, and mixed with oil of rofes. Change this twice a day, keeping the breaft very warm with flannel.*

If it be only a fmall indolent tumor, it would be better that nothing be done, as even the warmth of the above poultice, and the repeated application of flannel, have, by their ftimulus, fometimes converted fuch hardneffes into cancers, a fpecies of difeafes, of all others the moft to be dreaded ; whereas, by omitting the ufe of external means, fuch indurations have remained in an indolent ftate during the whole life of the perfon.

A Cancer in the Breaft.

No. 129. *Of thirteen years ftanding, was cured by frequently applying red poppy water, plantane and rofe water, mixt with honey of rofes. Afterwards the waters ufed alone perfected the cure.*

Of this extraordinary cure we have no evidence but Mr. Wefley's *ipfe dixit.*

No. 130. *Ufe the cold bath daily, (this has cured many.) This cured Mrs. Bates of Leicefterfhire, of a cancer in her breaft, a confumption, a fciatica, and rheumatifm, which fhe had near twenty years. She bathed daily for a month, and drank only water.*

We fhould be glad to be informed, in what part of Leicefterfhire Mrs. Bates lives ; it is a county of fome extent, and if the lady *really exifts* any where, it would have been proper to have given a more particular direction. We are induced to fay this, becaufe the relation

tion is too improbable to be credited by any perfons of common underftanding.

No. 132. *Rub the whole breaft morning and evening with fpirits of hartfhorn.*

Mr. W. appears to have no idea of the difference between fuch a tumour in the breaft, as is called by the furgeons a fcirrhous, and the exulceration termed a cancer. As I have obferved before (and I cannot help repeating it) it is not at all uncommon for women to have little fwellings arife on their breaft from various caufes, which, if let alone, hardly ever terminate ill either to the general health, or to the part affected. But if pretenders to medical knowledge or defigning quacks, advife hartfhorn, or other ftimulants, to be rubbed upon the part, with a view to difcufs fuch tumors, a greater fecretion of watery fluids is brought to the breaft from the ufe of fuch ftimuli; and thus what was at firft a very flight complaint, has been often converted into an incurable cancer. The writer declares with the greateft concern, that he has more than once feen in confequence of mal-practice, fuch an unhappy termination take place; and he moft earneftly advifes thofe who have any complaints of the breaft, to confult thofe who are poffeffed of fkill and humanity in the profeffion.

No. 135. *Take horfe fpurs, and dry them by the fire till they will beat to powder, fift and infufe two drachms in two quarts of ale; drink half a pint every fix hours; new milk warm.—It has cured many. Tried.*

No. 136. *Apply goofe dung and celandine beat well together, and fpread on a fine rag. It will both cleanfe and heal the fore.*

Mr. W. advifes *horfe fpurs* as an internal medicine, and *goofe dung* as an outward application; together with many other remedies for the cure of cancers,

equally

equally unaccountable. It is a melancholy truth, that ignorant men have always curatives in abundance for incurable complaints : as for the medical virtues of the many prescriptions advised by Mr. W. for can-cers, there can be little more objection to them, than to his powder of toad in the convulsive asthma.

A Cancer in the Mouth.

No. 141 to 150, Mr. W. has prescribed several external applications for the cure of cancers affecting the mouth ; and altho' one has the mark of *infalli-bility* affixed to it, and another the word *tried* ; not-withstanding all his boasted remedies, it will be highly prudent in so serious a complaint to apply to a good surgeon, who, it may reasonably be presumed, will make use of those applications which his expe-rience and judgment inform him are the most likely to prove successful.

I shall conclude my observations on cancers with one general remark ; which is, that interested and designing men have called every ulcer of *difficult cure*, which attacks the breast or mouth, a cancer ; and unfortunately the person so afflicted is unable to distinguish the one from the other. But it may safely be affirmed, that no real cancer was ever cured, it being a disease dependant on the laws of fermen-tation, and for that reason cannot be eradicated out of the constitution ; as one particle of cancerous matter remaining, is sufficient to renew all the ag-gravated symptoms of this horrid distemper. It is well known that experienced and able surgeons daily cure the worst ulcers ; and it is equally well known, that the humane and worthy part of that profession lament the many impostors who are every day starting up, to deceive the public with their pretended nostrums, and which too often, by their corrosive applications, increase the misery and hasten the death of the unhappy sufferer.

Children.

Children.

No. 157. *To prevent the rickets and weaknefs, dip them in cold water every morning till they are eight or nine months old; afterwards their hands and feet.*

The cold bath may not be improper for children, naturally of a good conftitution, though thefe feldom require its ufe; for in the cafes where this remedy is advifed, it is generally in confequence of fome difeafes, which have weakened and relaxed their tender frame, and therefore Mr. W. ought to have been a little more explicit; however, I fhall here take the liberty of giving a few hints relative to the ufe of the cold-bath, in fuch cafes, which if attended to, may be beneficial.

1ft. As the ftomach and bowels of young children are very apt to be difordered, it undoubtedly would be improper to ufe this remedy, when the child is affected with complaints of the firft paffages, as vomiting, purging, &c.

2dly. If any eruption fhould arife on the fkin, the cold-bath might prove a repellent, and therefore would have a tendency to be highly injurious; as fometimes by fuddenly ftriking in only a few pimples, an internal imflammation or fever has been brought on, the confequences of which have proved fatal.

3dly. If any fever fhould arife, whether from teething or any other caufe, it would be extremely dangerous to ufe the cold-bath.

After a child is eight or nine months old, Mr. W. reftrains the ufe of the cold-bath to the hands and feet only. If the child thrives from its being dipped in water, there can be no folid objection advanced to its being applied to the body univerfally after that age.

No

No roller should ever be put round their bodies, nor any stays used. Instead of them, when they are put into short petticoats, put a waistcoat under their frocks.

The easy method of dressing young children, is extremely well calculated for the promotion of health, and must be advantageous to the constitution ; but the public are indebted to the ingenious Dr. CADOGAN, for this improvement in the easy dressing of children : it is however not Mr. Wesley's method to acknowledge from whence he borrows his information, or whose words he makes use of. Whether justice be any part of *his* theological system or not, he has long been eminent for paying no attention to literary justice.

'Tis best to wean a child when seven months old.

It certainly would not be best. It would be much better to let the child have the breast two or three months longer; as no food can be substituted at that age so proper or so nourishing. The coagulable matter does not then abound too much, nor has the milk any pernicious properties, as I am convinced from experiment.

Let them go bare-footed and bare-headed, till they are three or four years old at least.

In many cases the going *bare-headed* at so early an age would be very improper, as there are not a few children who have not a sufficient quantity of osseous matter to prevent external injuries.

No child should touch any spirituous or fermented liquor, nor animal food, before two years old. Their drink should be water ; tea they should never taste till ten or twelve years old ; milk, milk-porridge and water-gruel are the proper breakfast for children.

These directions are undoubtedly very proper, and it would have been well if all Mr. Wesley's prescriptions

fcriptions had been equally innocent and unexceptionable.

Mr. W. has given fundry directions refpecting young children and their difeafes; but I muft do him the juftice to obferve, that he has never once recommended *Godfrey's cordial.* And, indeed, this is a very pernicious opiate, however frequently it may be adminiftered by ignorant nurfes. They give it to children for their own eafe, without confidering or underftanding its tendency. It has unqueftionably been productive of much mifchief; it tends to ruin the conftitutions of children, and the confequences of taking it have been much more frequently fatal than is generally apprehended. It is the earneft wifh of the writer, that this hint may be duly attended to by parents and thofe who have the care of young children. The ftate of our national population is at too low an ebb, for the lives of children to be facrificed to the ignorance of old women, or to the indolence of nurfes.

Chin Cough, or Hooping Cough.

No. 158. *Ufe the cold bath daily.*

I do not know upon what principle the cold bath can be advifed in this complaint. From the violent and long fits of coughing there is often much danger of fuffocation, and fometimes the agitation during the fits is fo violent as to rupture one or more of the blood-veffels of the lungs; fo that in every point of view this remedy, by loading the internal veffels *more* with blood, is likely to be highly injurious.

In defperate cafes change of air alone has cured.

Mr. W. after prefcribing feveral infignificant remedies, fays, " change of air alone has cured." In this direction he is perfectly right, but I would improve upon his advice, and earneftly recommend the
change

change of air at the very beginning of this very troublefome difeafe, as experience has convinced me that more real good may be done by this than by any other means.

I will here take the opportunity of making a few obfervations on the hooping cough, which may, perhaps, be not unworthy of attention.

1ft. I am clearly convinced, that this diforder arifes from infectious matter, and, like other difeafes, produced from a fimilar caufe, it has its beginning, progrefs, and decline; fo that when it is paft its acme, if an old woman happens to come in at the declenfion of this complaint, and advifes any thing ever fo abfurd, the cure is attributed to that; and indeed the fame reafoning holds good with regard to many of Mr. W.'s *infallibles* or *tried* remedies.

2dly. Medicine can do very little with regard to this difeafe, and yet it requires the attention of a good practitioner; for, if the violence of the cough produces *any inflammation of the breaft*, an occafional or even repeated bleeding will be undoubtedly proper; or if *coftive*, gentle laxatives will be highly neceffary; or, if the *phlegm fhould become too vifcid, and thrown up with difficulty*, expectorating medicines fhould be adminiftered.

3dly. *Affes Milk*, perfevered in for fix or eight weeks, has done effential fervice in this difeafe, as by its balfamic qualities, and the light nourifhment it affords, the juices are rendered mild and bland, which may often prevent the inflammatory ftate of the difeafe, or if it has come on, prevent its arifing to any confiderable degree; and thus exulceration of the lungs, or hectic fever, be kept off; fo that the patient may get through this troublefome and tedious complaint without the danger commonly attending it.

Cholera

No. 166. *Drink two or three quarts of cold water, if strong ; of warm water, if weak.*

No. 167. *Drink a draught of vinegar and water.*

No. 168. *Boil a chicken an hour in two gallons of water, and drink of this till the vomiting ceases.*

These are insignificant remedies, prescribed by Mr. W. in a very serious disease ; but it may be presumed that the pain, uneasiness, &c. will generally oblige those who are attacked with this complaint, to have recourse to proper advice; which, indeed, is absolutely necessary ; for if this violent affection of the stomach and bowels is not removed within forty-eight hours, it brings on such a universal debility of the whole system, as generally proves fatal.

No. 169. *Take six grains of laudanum.*

As Mr. W. uses the word *grains*, he must undoubtedly mean *solid opium*. Now, opium is one of those *Herculean medicines*, which he so pathetically dissuades his readers from the use of, in his preface ; and yet he here prescribes a dose of this powerful narcotic, which would, in all probability, cause ninety-nine persons out of a hundred *to sleep for ever*. In disorders in common, the Faculty seldom admister above one grain, and scarcely ever exceed two grains. It is true, that Mr. BROMFIELD and Mr. POTT have lately given this drug in larger doses, in extraordinary chirurgical cases with great success ; but I believe these gentlemen have never yet ventured to direct, at the first or second exhibition of this medicine, six grains for a dose ; but a remedy that would be safe and useful in the hands of these skilful and eminent practitioners, becomes a dangerous weapon in the hands of the ignorant and unskilful.

The

The Colic, (in the Fit.)

No. 180. *Drink a pint of cold water.—Tried.*

No. 181. *A quart of warm water.—Tried.*

No. 182. *As largely as poſſible of warm tar-water.*

No. 183. *Or a pint of water in which a red hot flint is quenched.*

This inimitable profeſſor of phyſic, preſcribes *hot* and *cold* remedies in the ſame breath. A pint of cold water, he ſays, is a *tried* remedy for the colic ; but if you are not ſatisfied with that, *a quart of warm water* is another *tried* remedy for the ſame diſorder, and will do full as well. And if you ſhould not reliſh either of theſe curious preſcriptions, he adviſes you to drink *as largely as poſſible of warm tar water*; or that you may have another choice, *a pint of water in which a red hot flint is quenched.* Two of theſe watery remedies are *tried*, and two, it ſeems, are *untried*; but I am of opinion they are equally infallible ; though if any one deſerves the preference, it is the tar water.

No. 186. *Take thirty drops of ſpirits of turpentine in a glaſs of water.*

If the perſon affected with the colic, is of a ſtrong, or an inflammatory habit ; or if there be any obſtruction in the bowels, this remedy muſt prove exceedingly injurious.

No. 187. *Or from two ſcruples to half a dram of yellow peel of orange, powdered, in a glaſs of water.*

As half a dram is 30 grains, and two ſcruples 40 grains, would it not have been a little more methodical, to have ordered the doſe of orange peel to be taken, from half a dram to two ſcruples ?

No. 188.

No. 188. *Beat together into a cake, one part of stoned raisins of the sun, and three parts of juniper berries; eat more or less, according to the pain.*

Is the patient to eat an ounce, or a pound?

No. 189. *Take from 30 to 60 drops of oil of aniseed, on a lump of sugar.*

As this essential oil, like all others, possesses a considerable degree of stimulus, it ought to be taken with much caution, of which Mr. W. appears to have no conception; but with a want of precision which seems natural to him, he advises 30 or 60 drops of oil of aniseed indiscriminately.

Bilious Colic.

No. 194. *Give a spoonful of sweet oil every hour. This has cured one judged at the point of death.*

If the experiment has been made only upon one person " at the point of death," I hope, in so violent a disease, that other aids will be called in to prevent the patient being brought into so much danger. In general, a good practitioner finds this disease yields very readily; so that such imminent danger may be easily avoided.

No. 197. *Mrs. Watts, by using the cold bath two-and-twenty times in a month, was entirely cured of an hysteric colic, fits, and convulsive motions, continual sweatings and vomiting, wandering pains in her limbs and head, with total loss of appetite.*

This recovery of Mrs. Watts is, unquestionably, a very surprising one, and would deserve little credit, were it not for the very satisfactory manner in which it is authenticated. Mrs. Watts, by the use of the cold bath, was entirely cured, Mr. Wesley informs us, not only of an *hysteric colic, continual sweatings and vomiting, wandering pains in her limbs*

E

and

and head, and also of a total loss of appetite into the bargain. All this is sufficiently marvellous; but perhaps an inquisitive reader might be tempted to enquire who Mrs. Watts is, and where she dwells, whether in England, Scotland, Ireland, or America; but these trifling circumstances, the sagacious Mr. W. chuses to bury in profound silence.

No. 198. *In the fit, drink half a pint of water, with a little wheat flour.*

No. 199. *Warm lemonade.*

No. 200. *A glass full of vinegar.*

An hysteric colic is the general consequence of a weak state of the stomach and bowels, in which acidity is mostly predominant; and therefore the glass of vinegar or lemonade must, in every point of view, be highly injurious.

A Nervous Colic.

No. 204. *Use the cold bath daily, for a month:*

In this disease, Mr. W. advises the cold bath indiscriminately to be used for a month; but there are many circumstances, with regard to the nervous colic, which may render this remedy extremely improper; not to mention the length of time during which it would be necessary to persevere in the use of it.

No. 205. *Take quicksilver and aqua sulphurata daily, for a month.*

Upon what principle quicksilver and aqua sulphurata (which is a diluted vitriolic acid) is to cure the nervous colic, is very difficult to ascertain; to me, these remedies do not seem to promise any good effects; but, on the contrary, in a weak state of the bowels, would rather exasperate the disease.

In a note, Mr. W. says this colic is frequently " termed the dry belly-ach." I believe these two dif-.

orders

orders of the bowels, can only be called the same diseases, by those persons, who, like Mr. W. are ignorant of the nature of the symptoms which constitute the two complaints ; as it is generally understood by good practitioners, that the hysteric and nervous colic are one and the same disease ; and not to be confounded with the colica pictonum, or dry bellyach.

Colic from the Fumes of Lead, White Lead, Verdigris, &c.

No. 206. *In the fit, drink fresh melted butter, and then vomit with warm water.*

No. 208. *To prevent or cure ; breakfast daily on fat broth, and use oil of sweet almonds frequently and largely.*

This is the disorder that is generally termed the dry belly-ach, and very properly so ; but it is hardly necessary to make any remarks on this colic, as the excessive pain, and other symptoms attending it, will generally oblige the afflicted to apply for assistance ; which will be acting with much more prudence, than placing any dependance on Mr. W.'s very futile prescriptions.

In all the various colics, Mr. W. has not advised one purging or laxative remedy ; (indeed, in the bilious colic, he has prescribed a spoonful of oil) but happily, people in general, in these disorders of the stomach and bowels, take of themselves, or are advised to take, tincture of rhubarb, Daffey's elixir, or some other remedy of a similar kind ; which, by unloading the first passages of indigestible and other offending substances, will often, in a very short time, remove such colic complaints ; but when the bowels have been emptied, and the pain or other symptoms continue, or should return, then the warm and stomachic me-

E 2   dicines

dicines may generally be adminiftered with fafety, and will moftly carry off the complaint.

It may alfo be obferved that colics attack the ftrong as well as the weak; and in thofe cafes where there is a fixed pain in one part of the belly, the pulfe hard and frequent, with other inflammatory fymptoms, it will be highly neceffary to ufe the lancet as foon as poffible; becaufe an inflammation is then forming in a fmall part of the bowels, which, if not timely removed, may in twenty-four or forty-eight hours, terminate in a mortification of the part affected; and if fpirituous remedies fhould be imprudently ufed to remove this pain, having a natural tendency to increafe the fymptoms of the difeafe, they may fometimes kill in a very few hours. Many lives have been undoubtedly loft in this manner, and therefore it may often be highly dangerous for perfons in colic diforders to have recourfe to Mr. W's remedies of fpirits of turpentine, oil of anifeed; or to double diftilled waters, peppermint, or any other kind of fpirituous cordial, which many good women in the country, whofe medical fkill may be fuppofed equal to that of Mr. Wefley, are too apt to recommend in fuch cafes.

A Confumption.

No. 211. *A beginning confumption was cured by drinking decoction of guiacum, morning and evening, for fifteen days, (without fweating) with a light diet.*

In every view in which this medicine can be confidered, it appears abfolutely improper, as the guiacum owes its medical virtues to a ftimulating effential oil, which would moft probably heat the fyftem, and thus the remedy prefcribed might convert a beginning confumption into a confirmed hectic. Mr. W. alfo, with his *ufual inaccuracy,* only advifes a decoction of the guiacum wood, but he does not give his readers any

infor-

information refpecting the quantity of guiacum to be
ufed, or what the proportion fhould be between that
and the watery fluid, nor whether the confumptive
patient is to take it in half pints or in gallons.

No. 212. *Cold bathing has cured many deep confumptions. Tried.*

Cold bathing, Mr. W. informs us, is a tried remedy for confumptions, and " has cured many deep
confumptions." As, according to him, this is almoſt
a fpecific in this difeafe which is fo peculiar to this
country, and which phyficians generally find fo
extremely difficult of cure ; it would have been well
if Mr. W. had been a little more particular on this
head, and informed his readers whether the cold bath
cured incipient confumptions only, or in what ſtage
of the difeafe, it effected the extraordinary recovery
he fpeaks of. In fo fatal a difeafe it certainly would
have been worth while, if he really had poffeffed that
regard for the lives of his fellow creatures which he
affects, to have informed us what thofe kinds of confumptions were, in which the cold bath proved fo fur-
prifing a fpecific.

Cold bathing is certainly a very eafy remedy for
the cure of a confumption ; but Mr. W. who deals
much in wonderful recoveries, has another remedy
for this dangerous difeafe ſtill more eafy than the
former. He acquaints us (at No. 213) that in three
months time, a perfon in a *deep confumption* was perfectly reſtored by drinking *nothing but water, and
eating nothing but water-gruel, without falt or fugar !*
This was truly marvellous, and we fhould have been
glad to have been informed where the man lives,
what his name is, or to have had the ſtory in fome
degree authenticated. But perhaps Mr. W's affertion is fufficient, it may be fo in the neighbourhood
of Moorfields ; but we believe his *veracity* is not fufficiently *eſtablifhed at Briſtol* for his *ipfe dixit* to pafs
there for unqueſtionable truth.

No.

No. 221. *Every morning cut up a little turf of fresh earth; and lying down, breath into the hole for a quarter of an hour.—I have known a deep consumption cured thus.*

Here is another of Mr. W.'s remedies for a consumption, which needs only be mentioned to excite the readers risibility. It is a recipe indeed truly worthy the acute genius of the author of Primitive Physic.

No. 227. *Take in for a quarter of an hour, morning and evening, the steam of white rosin and bees wax, boiling on a hot fire-shovel.—This has cured one who was in the third stage of a consumption.*

There is no end to the discoveries of Mr. W. in the cure of consumptions: the above is as extraordinary a remedy as the recovery is astonishing; for when a person is unhappily arrived at the third stage of a consumption, the lungs are generally so deeply ulcerated, that the most able practitioners cease to have any hopes, for the disease is so far advanced as to be past the power of medicine. But Mr. W. can even perform cures then, and that by *rosin* and *bees-wax boiling on a hot fire-shovel.* It is however to be regretted, that we are not informed of the name of the person thus surprisingly cured, and of the place of his abode. But Mr. W.'s prudence, or art, or effrontery, is superior to that of common quacks. They generally pretend at least to give some information where the persons they have recovered are to be found; but Mr. W. is above every thing of this kind. He says, that by his recipes great cures have been performed; and to enquire of whom, and how the facts are ascertained, is an impertinence. But he should remember, that all the people of England are not votaries to implicit faith, however strongly it may actuate the patient hearers at the Foundery.

To

To ftrengthen the body, take falt-petre half a drachm,
falt of fteel fifteen grains, in a quarter of a pint of wa-
ter.   Add two ounces of the beft brandy, and fweeten it
with loaf fugar ; drink two fpoonfuls of this about eleven
in the morning, and at five in the afternoon, wafhing it
down with a difh of fage tea.   This mixture may be
repeated twice or thrice.

This ftrengthening prefcription is an unqueftionable proof that Mr. W. is totally ignorant of *double*
elective attractions, in other words, that by diffolving
a neutral and metalline falt in water, two new com-
pounds are formed ; fo that the weak patient prefcribed for, is neither taking falt-petre nor falt of fteel,
as a ftrengthening remedy ; but new combinations
are produced by the folution, which are totally different in their medical effects on the human body, as
well as in their chemical properties.

This inftance, among many others, may ferve to
fhew how unfafe it is for mankind to follow impli-
citly the prefcriptions of ignorant pretenders, and
thofe who, like Mr. W. are deftitute of chemical or
medical knowledge.

In the 10th page of his preface, he complains
heavily of the phyficians for introducing " into
" practice abundance of compound medicines, con-
" fifting of fo many ingredients, that it was fcarce
" poffible for common people to know which it was
" that performed the cure ;" and alfo " chymicals,
" fuch as they neither had fkill, nor fortune, nor
" time to prepare.   Yea, and dangerous ones, fuch
" as they could not ufe, without hazarding life, but
" by the advice of a phyfician."   And here this
very fcrupulous and affectedly cautious gentleman
recommends a neutral falt, the elements of which
are fixed vegetable alkali, and the nitrous acid, and
alfo a metalline combination, whofe elements are iron
and the vitriolic acid : two chemical compound medicines, the nature and effects of which it is manifeft
he

he is totally ignorant of. The writer thinks he
may with the ftricteft juftice apply Mr. W.'s own
words to himfelf*, with a fmall variation; "How
" many inconveniencies" may have been occafioned
by fuch ignorant prefcribers! "How many conftitu-
" tions" may have been ruined! "How many va-
" luable lives loft."

I fhall conclude the whole of what has been faid
relative to confumptions, with advifing thofe who
have an obftinate cough, which occafions a great deal
of vifcid phlegm to be thrown up, and is attended with
a fudden emaciation, and other incipient fymptoms
of a hectic; to apply immediately for proper affiftance. Perfons in this ftate are generally able to
walk about, fo that they do not confider the danger of their prefent fituation; but the writer affures them thefe are the marks of a beginning confumption, therefore immediate recourfe
fhould be had to thofe of fkill in the profeffion.
And this is attainable even by the poor, as phyficians
are always ready to do every kind office to the indigent; and in the metropolis particularly, which is
fo honourably diftinguifhed by the number of its
charitable inftitutions, the pooreft perfon need not be
at a lofs to procure proper advice. By attending
to the difeafe at firft, *hundreds* might every year be
reftored to health, who, if the diforder be once
fixed, would be carried off by confumptions. I
do moft earneftly intreat perfons of all ranks to
have early advice in fuch affections of the breaft,
and by no means to neglect the cough, &c. which
may be readily cured at firft; but when continued
fome time, may become an incurable difeafe. The author is thoroughly convinced, from experience, that
if proper precautions were ufed at the beginning of
confumptions, not one in a hundred would die of a

* See Preface, page xxvii.

dif-

difeafe, which, through neglect and inattention, carries off vaft numbers every year.

Convulfions.

No. 232. *Ufe the cold bath.*

No. 233. *Take a tea-fpoonful of valerian root every
evening.*

No. 234. *Half a drachm of mifleto, powdered, every
fix hours.*

Convulfions in Children.

No. 235. *Scrape piony roots frefh digged; apply
what you have fcraped to the foles of the feet. It helps
immediately. Tried.*

Convulfions in the Bowels of Children.

No. 236. *Give a child a quarter old, a fpoonful of
the juice of pellitory of the wall, two or three times a
day. It goes through at once, but purges no more.*

As convulfions are not a difeafe, but generally the
confequences of other diforders; as they arife in all
habits of body, the weak, the ftrong, and the ple-
thoric, being fubject to them; it was neceffary that
Mr. W. fhould have attended a little to thefe circumftances: but as convulfions are a very frightful and
alarming appearance, there are very few who will
attend to his prefcriptions, at leaft they will not, if
they are under the influence of prudence.

A Cough.

No. 350. *Every cough is a dry cough at firft. As
long as it continues fo, it may be cured by chewing immediately after you cough, the quantity of a pepper corn of*

F   *Peruvian*

Peruvian bark. Swallow your spittle as long as it is bitter, and then spit out the wood: if you cough again, do this again. It very seldom fails to cure any dry cough. I earnestly desire every one who has any regard for his health to try this within 24 hours, after he first perceives a cough.

The bark is one of those Herculean remedies, against the use of which Mr. W. dissuades his readers, and which he says *are too strong for common men to grapple with.* He says also, that they are *edged tools;* but that the physicians *have not yet taught them to wound at a distance:* and he adds, that *honest men are under no necessity of touching them, or coming within their reach.* And yet he recommends this formidable remedy to every person affected with a cough. Is there any consistency in this? But he has long been distinguished for his variableness and inconsistency. But however contradictory and absurd his recipes are, it is one consolation, that like the four Herculean medicines, *honest men are under no necessity of touching them, or coming within their reach.* *

From 251 to 272, Mr. Wesley prescribes many remedies for coughs, but they are unworthy of attention; however, I shall here take the liberty to observe, that a cough is only the symptom of an asthma, catarrh, peripneumony, pleurisies, &c. and a good practitioner will consider what the diseases are, which occasion the cough, and will prescribe accordingly.

To cause an easy Delivery.

No. 296. Peel, slice, and fry a large white onion, in two or three spoonfuls of the best oil, till it is tender, boil this with half a glass of water; strain it, and drink it in the morning fasting, for two or three weeks before the time of child birth.

* Wesley's Preface, p. 24.

Our

. Our fagacious author here prefcribes *a fliced onion, firft fried, and afterwards boiled*, to caufe an eafy delivery. It may be hoped that all the practitioners in midwifery, male and female, in this kingdom, will pay due regard to this ingenious prefcription.

A Diabetes.

No. 297. *Drink wine boiled with ginger, as much and as often as your ftrength will bear.*

Here is a very ftrange remedy prefcribed for the diabetes, and no regard whatever paid to the quantity of wine to be ufed, or the dofes of ginger to be taken ; furely, in prefcribing wine and ginger as a medicine, the dofe, and times of exhibition, were circumftances worthy of fome little attention.

No. 299. *Infufe half an ounce of cantharides in a pound of elixir of vitriol. Give from 15 to 30 drops, or even 40 drops, in Briftol water, twice or thrice a day.*

As in this difeafe, the nutritious and balfamic parts of the blood are carried of by the kidneys, fo that great weaknefs, emaciation, and hectic fever will come on, if the diforder is not very foon removed; it is to be hoped that every perfon affected with the diabetes, will have recourfe to proper advice, on its firft approaches.

. Mr. W.'s prefcription of cantharides in this difeafe, is equally abfurd and dangerous ; and the writer hopes no perfon will be fo credulous or incautious as to make ufe of it.

The Dropfy.

No. 300 to 323. Mr. Wefley gives 23 prefcriptions for the cure of dropfies, and fays fuch extraordinary things of fome of them, that it were to be wifhed the facts had been better authenticated.

F 2   The

The Rev. Mr. GRANGER, in his ingenious biogra-cal work, fays of the *Primitive Phyfic*, that "this book, by the help of the title, hath had a good run among the Methodifts, whofe faith, co-operating with nature, frequently made them whole, when Mr. W. had the credit of the cure."

Drowned.

Mr. Wefley recommends, from Dr. Tiffot, that the trunk of the body of a drowned perfon, fhould be rubbed all over with falt. It is not neceffary to make any remarks on this ; but I fhall here take the liberty to obferve, that as the fociety lately eftablifh-ed in London, for the recovery of perfons apparent-ly drowned, &c. (an inftitution which my worthy and ingenious friend, Dr. COGAN, and myfelf, affifted by many refpectable gentlemen, have been happily inftrumental in introducing into this kingdom) have given their methods of treatment to the public, and which, in the fpace of eighteen months, have been the happy means of RESTORING FORTY-ONE PERSONS TO LIFE ; it may reafonably be hoped, that thefe methods, the efficacy of which has been demon-ftrated by unqueftionable facts, will be duly at-in all cafes of this nature. Thofe who wifh for further information on this fubject, may meet with it in the PLAN and REPORTS of the Society, printed in the prefent year.

On Fevers.

It is with regret that the writer trefpaffes on the patience of his readers, by troubling them with Mr. W's abfurd divifion of fevers, and his very incon-fiftent remedies for a difeafe that carries off three parts of the human fpecies ; but it is neceffary, in or-der to give them the more thorough conviction, that the

the *Primitive Phyfic* is a publication calculated to lead thofe who rely upon it, to trifle with their lives, in the moft dangerous and alarming difeafes.

No. 402. Toafted bread and water can do no hurt in a fever; and it may, therefore, very fafely be given, either in a *dry heat*, or a *moift heat*, to adopt the curious language of our profound practitioner.

It is not eafy to meet with any quack, even the moft affuming, who profeffes to cure difeafes with more facility than Mr. W. If his directions are followed, diforders, of the moft dangerous kind, difappear, as at the touch of the magician's wand. He cures a burning fever in an hour. What were *Hippocrates* or *Galen*, compared to John Wefley! *Stamp* (fays he, 408) *a handful of leaves of woodbine; put fair water into it, and ufe it cold, as a clyfter. It* COMMONLY *cures in an hour.* A more expeditious remedy need fcarcely be wifhed for; much is it to be regretted, that its efficacy is not fomewhat better authenticated! But it is to be feared, that Mr. W's faith in this remedy is only founded, as implicit faith generally is, upon ignorance. He feems to have no idea, that the burning heat in a fever, will frequently abate on a fudden, and go off in an hour, and yet return again in a few hours, with equal violence; and what he fuppofes to have been a cure, could be only a temporary abatement; and even this is very unlikely to be procured in a burning fever, by a cold clyfter.

Mr. W. may perhaps imagine, that when what he recommends as remedies are not manifeftly pernicious, if he does no good, at leaft he does no harm. But this in many cafes will be a moft *egregious miftake*, and of this the prefent prefcription is an inftance. He recommends (No. 412) *thin water-gruel*, or *boiled milk and water*, in a *hectic fever*. As fimple drinks, thefe can do no harm: but thofe who are led by their confidence in Mr. W's opinion, to rely upon thefe as

probable

probable remedies for the cure of a hectic fever, may sustain an irreparable injury. By expecting relief from those things which cannot cure them, they are prevented from having recourse to those remedies which, if they had been taken in time, would have been efficacious. An hectic fever is a disease that requires judicious and attentive practice, and which seldom has a fatal tendency, if the sick person is not led by such dablers in physic as Mr. W. to trifle too long with a disorder which, after a certain length of time, becomes incurable.

Of all the diseases to which human beings are subject, there is none which carries off so many as fevers. But Mr. W. is possessed of a remedy for them all, which is equally easy and infallible. *Plunging in cold water*, he says, (No. 413) is a *safe and a sure remedy in any fever*: and he even recommends this in a high fever, attended with a *delirium* and a *vigilia*, which are the most dangerous symptoms with which a fever can possibly be attended. It would be a happy circumstance if this remedy were as certain and as safe as Mr. W. represents it; but in this case his mere assertion is surely not sufficient, unless among the most credulous of his followers; and, unfortunately, he has not supported what he has advanced by any facts, or by any kind of evidence.

The next disease for the cure of which Mr. W. attempts to prescribe, is an *ague*, which he calls an *intermitting fever*, each fit of which is preceded by a cold shivering, and goes off in a sweat: for this he prescribes FORTY REMEDIES. It might have been presumed that these would have been sufficient; at least by a man who has said, " Experience shews " that *one* thing will cure most disorders, at least as " well as *twenty* put together. Then why do you " add the other nineteen." But though Mr. W. is offended that many remedies should be prescribed by other people, he himself does not think *forty* sufficient.

ficient. And therefore in the 8oth page of his book he gives *five* more remedies for an intermitting fever, which are ſo different from thoſe he preſcribed for an ague, that he ſeems not to have known that they were the ſame diſeaſe, though he had ſaid they were in the firſt page of his pamphlet. But Mr. W. is too ſtrongly characteriſed by inconſiſtency, for any thing of this kind to excite our aſtoniſhment.

A Nervous Fever.

No. 423. *Drink every night a tea-ſpoonful of cream of tartar, boiled in half a pint of milk.*

This preſcription is nothing more than whey, and is the only one ſet down for the cure of this fever ; and it is mere trifling with the patient in a diſeaſe which is occaſioned by a great debility of the vital powers ; ſo that a crem of tartar drink cannot poſſibly be of the leaſt ſervice, but will generally be highly injurious.

A Raſh Fever.

No. 424. *Drink every hour a ſpoonful of juice of ground-ivy. It cures in 24 hours. Uſe the decoꞒion when you have not the juice.*

In a raſh fever, as it is termed by Mr. W. and other ignorant people, he recommends that the patient ſhould drink ground-ivy juice or decoꞒion. This preſcription is equally bold and unſupported ; it is entirely inapplicable to the diſeaſe, and none but the weak and credulous will expeꞒ any relief from it.

A ſlow Fever.

No. 425. *Uſe the cold bath for two or three weeks, daily.*

A nervous and ſlow fever, is generally allowed by practitioners to be one and the ſame diſeaſe; but we muſt not expeꞒ the author of the *Primitive Phyſic* to

entertain

entertain the fame ideas as the gentlemen of the Faculty; for, in the nervous fever, his remedy is cream of tartar and milk : in the flow fever, no internal remedy whatever is advifed, but he recommends, daily, the ufe cf the cold bath, for two or three weeks. It is too true, that the nervous, or flow fever, is apt to continue feveral weeks, and the fymptoms often become fo irregular, that it requires the utmoft attention to prefcribe fo as to give the remedies their greateft efficacy; but Mr. W. with his ufual indifference, indifcriminately orders cream of tartar and the cold bath, without paying the leaft regard to the different circumftances of this diforder, of which there are hardly two cafes alike.

As the author did not fit down, merely with a view to expofe the errors and abfurdities of Mr. W.'s performance, but alfo with a defign to offer his readers fuch obfervations as fhould occur to him, which might be of a beneficial tendency, he will here take the liberty of making a few general remarks, relative to Mr. W.'s method of claffing fevers, and the mode of treatment recommended by him, in fo violent and dangerous a difeafe.

Mr. Wefley claffes fevers under the following heads : *a fever, a burning fever, an acute fever, a continual fever, a hectic fever, an intermitting fever, a fever with pains in the limbs, a nervous fever, a rafh fever,* and laftly, *a flow fever.*

Upon which it may be obferved, that Mr. W. has read or underftood little of this fubject, to confider a burning fever, an acute fever, a continual fever, and a fever with pains in the limbs, as different kinds of fevers. My experience and obfervation have convinced me, that what are here fuppofed to be four different kinds of fevers, are one and the fame difeafe. As for inftance, with refpect to what is called a *burning fever*; is there not more or lefs heat, in general, in fevers? And is it not abfurd, that, becaufe

caufe the fever fhould be high at one time, and the
heat then increafed, that therefore it fhould be fpo-
ken of as a peculiar difeafe ?

2dly, As to what is called an *acute fever ;* as all
fevers are univerfally allowed to be acute difeafes,
there can be no ufe or meaning in this difcriminating
term.

3dly, As to the phrafe *continual fever* ; are not all
fevers continual, except intermittents ? And what
then is the meaning of this diftinction ?

4thly, As to what is termed *a fever with pains in
the limbs* ; in the very fame fever, are not the different parts of the body, varioufly affected with the difeafe ; fometimes pain in the head, fometimes pain in
the limbs ? If, then, we muft have a new name for
every accidental fymptom that arifes, we might have
as many claffes of fevers, as there are days in the
year.

As to the *rafh fever,* Mr. W. does not inform us
whether he means a miliary, a fpotted, or a petechi-
al fever, but advifes the patient to drink the *juices of
ground-ivy,* a medicine of no real virtue ; fo that it is
directing the fick to trifle in a difeafe of a very dangerous nature, and which requires the fkill of a fagacious practitioner.

It fhould alfo be obferved, that Mr. W. has not,
in the variety of fevers he has enumerated, given the
fymptoms of any one of them ; fo that the fick, or
their friends, are left to guefs at what kind it is, with
which they are attacked ; and when fuch a vague
method of proceeding is to be adopted, is it not ten
to one that they miftake the fever, and thus endanger
the life of the patient ?

Thofe who are called to the affiftance of perfons
afflicted with fevers, ought to attend carefully to the
fymptoms and progrefs of the difeafe. It is a rule
with me, in cafes of this nature, to vifit my patients

G   twice

twice a day ; and I often find, that within the com-
pafs of a few hours, there is fo great a change in the
difeafe, that the plan I had formed in the morning, is
abfolutely improper in the evening. Now, how is it
poffible, that juftice can be done to the afflicted, in a
diftemper which carries off fuch great numbers of the
human fpecies, without a knowledge of the caufes,
progrefs, and termination of difeafes. It is *this*, and
this only, that can lead to a rational and confcientious
mode of treatment ; and whoever prefcribes in acute
difeafes, when deftitute of this knowledge, whether
Mr. W. or any other Quack, will be in great danger
of violating the eighth commandment. Tho' the
moft judicious practitioner cannot always cure fevers, yet it is a fortunate circumftance for the patient,
when he is fo happy as to be attended by a careful
obferver of nature, and of the operation of his remedies.

When perfons are taken ill with fevers, apothecaries are generally fent for before a phyfician, and
therefore they ought to be well acquainted with the
duties of their profeffion. And when the apothecary is fent for in fuch a cafe, if he be poffeffed of
fkill and integrity (as many, it is prefumed are,
notwithstanding Mr. W's infinuations) he will not
only confider the fymptons of the difeafe, but the
ftate of the conftitution, and thereby be led to a
proper mode of treatment.

Among all Mr. Wefley's remedies for fevers,
bleeding is never once advifed to lower the action of
the veffels, which is exceedingly neceffary when the
pulfe is hard, full, or ftrong, and there are other fymptoms of inflammation in the habit; nor does he once
advife an *emitic* or a *purgative* at the beginning of fevers, altho' there may be fymptoms indicating their
their ufe in the ftrongeft manner, and caufed by ob-
noxious matters in the firft paffages; the removal
of

of which, early in the difeafe, will often caufe the fever to terminate in two or three days, when it would otherwife have run on for as many weeks.

It is of great importance that all difeafes, and thofe of the acute kind in particular, fhould be taken care of in the beginning. A difeafe may be almoft incurable after it has continued for fome days, which might have been eafily cured at the firft attack. When a fever, or any internal inflammation is neglected for a day or two, or fuch improper and futile remedies ufed as are advifed by Mr. W, or fimilar pretenders to phyfic, the former will frequently run on for many days, and the latter terminate in fuppuration of the part, and probably both in their confequences prove fatal.

Mr. Wefley, like many others who have not paid a due attention to the hiftory and progrefs of difeafes, often prefcribes only for *fymptoms*. Thus, he has his *cold bath for delirium and vigilia*, his *lambs-lungs alfo for delirium, hartfhorn drops for a fever with pain in the limbs, &c.* And this leads me to remark, that I have been thoroughly convinced from feventeen years experience, that prefcribing to particular fymptoms, is a moft dangerous mode of practice. There are fome who will prefcribe for the head-ach, others for pains in the limbs, &c. not reflecting that thefe are only fymptoms of the difeafe called a fever; but becaufe they are fymptoms which give pain and uneafinefs to the fick, they are particularly noticed by them. But it fhould be remembered, that thofe fymptoms which *give no pain*, are the moft dangerous part of the difeafe; fuch as the appearance of the eye, which fhews the ftate of the brain; the *pulfe* which fhews the ftate of the vafcular fyftem; and the *tongue*, and *urine* which fhews the ftate of the blood and the fecretions. And when the fever goes off, thefe, as a part of the difeafe, will naturally go off alfo: but remedies which are prefcribed merely for the removal of painful fymptoms, are by no means

means the way to effect a radical cure of the difeafe termed a fever.

In the 24th page of his preface, Mr. Wefley intimates, that the art has been difcovered of preparing quickfilver in fuch a manner, "as to make it the " moft deadly of all poifons," and he cautions his readers againft it with great vehemence; but, notwithftanding this, Mr. W. advifes (No. 426) for the cure of a fiftula, a folution of corrofive fublimate in fpring-water; and he fays alfo, under this head, that the fame medicine will, in forty days, *cure any cancer*, or *any running fore, or king's evil, broken or unbroken.* His averfion to mercurials has not prevented him from here recommending one of the moft active mercurial preparations, and one which requires the greateft fkill to adminifter properly in difeafes, and alfo the greateft care with refpect to circumftances. And indeed no perfon can, with any prudence or fafety, take this active medicine under no better directions than thofe given by Mr. W.

Flegm.

No. 430. *To prevent or cure, take a fpoonful of warm water the firft thing in the morning.*

FLEGM. I confefs, that I was for fome time at a lofs to know what the learned author meant by this word, which I do not remember ever to have met with before, and had fearched many dictionaries without finding any fuch term. Neither did the ingenious prefcription to prevent or cure *flegm* afford me any elucidation on this intricate fubject; for I imagine a fpoonful of warm water to be as much a panacea, or univerfal remedy, as it is a fpecific in any particular diforder. At laft I conjectured *phlegm* might be meant, and that Mr. W. had pillaged fome old manufcript of his grand-mothers for a recipe to

pre-

prevent or cure flegm ; and that we might be certain of its authenticity, he had faithfully copied the old lady's orthography.

A Flux.

No. 433. *Ufe the cold bath daily, and drink a draught of water from the fpring.*

The cold bath is recommended for the cure of a flux, which muft be exceedingly improper, as by fuddenly contracting the fuperficies of the body, a load of fluids will be determined to the interior parts, fo that the quick and great diftenfion of the interior veffels will have a tendency rather to increafe the flux than to cure it.

A Bloody Flux.

No. 445. *Drink cold water as largely as poffible, till the flux ftops.*

Mr. W. advifes a perfon affected with the bloody flux to " drink cold water ;" but what is more extraordinary than the remedy, he recommends that nothing elfe fhould be taken " till the flux ftops." Here a very ineffectual remedy is prefcribed for a very dangerous difeafe, and if the patient is to take nothing elfe, he is configned over to certain death, unlefs the flux ftops of itfelf ; for no reafonable man can be of opinion that it will ever be ftopt merely by drinking cold water.

No. 447. *Take a large apple, and at the top pick out all the core, and fill up the place with a piece of honey-comb (the honey being ftrained out) ; roaft the apple in embers, and eat it, and this will ftop the flux immediately.*

It were to be wifhed, that there was fome evidence of the efficacy of this extraordinary remedy for the bloody flux ; for till this is produced, thofe who
expect

expect a cure from it muſt be poſſeſſed of more credulity than underſtanding.

Mr. W. ſays, that *powdered root of gladwin is juſt
as good as rhubarb* in moſt caſes. But the ſuperior
efficacy of rhubarb has been ſo well aſcertained, that
this aſſertion would not have been made by any man,
who was acquainted with the medical principles of
theſe medicines.

No. 454. *A perſon was cured in one day by feeding
on rice-milk, and ſitting a quarter of an hour in a
ſhallow tub, having in it warm water three inches
deep.*

This rice-milk, ſhallow tub, and warm water preſcription, is a very important one; but the good
women who attend on this occaſion muſt take eſpe-
cial care, that the water is exactly *three inches deep*,
and that *the tub* is as *ſhallow* as the preſcription.

To prevent (or ſtop a beginning) Gangrene.

No. 455. *Foment continually with vinegar, in which
droſs of iron (either ſparks or clinkers) has been boiled.*

As a gangrene is an incipient mortification, which
if it once ſpreads to a vital part, generally deſtroys,
every honeſt and humane practitioner is always
alarmed at the ſtate of his patient; and if he be in
indigent circumſtances adviſes immediate recourſe to
hoſpital aſſiſtance; if otherwiſe, calls in a good
ſurgeon and phyſician, to ſtop, if poſſible, a diſeaſe
ſo dreadful in its conſequences. Mr. W. however,
contents himſelf with ordering an inſignificant external application. But the writer of this moſt ear-
neſtly recommends, if there be any ſuſpicion of a
gangrene attacking any part, that the beſt aſſiſtance
may be immediately applied for, as the delay of half
an hour may prove the death of the perſon.

Mr. W.

Mr. W. prefcribes no internal medicines for the removal of fo ferious a difeafe as a gangrene, but indeed we muft not wonder that he does not order the bark, as in his preface, page 24, he fays: that it is one of the Herculean remedies, *far too ftrong for common men to grapple with. How many fatal effects have thefe* (he includes, *antimony, opium, fteel and quickfilver) produced even in the hands of no ordinary phyfician.*

But to enable my readers to judge properly on this fubject, I fhall take the liberty of making a fhort extract from that ingenious and fkilful furgeon, Mr. POTT. He fays, " the powers and virtues of the *bark* " are known to almoft every practitioner in phyfick " and furgery. Among the many cafes in which its " merit is particularly and juftly celebrated, are the " diftempers called *gangrene and mortification;* its " general power of ftopping the one and refifting " the other, have made no inconfiderable addition " to the fuccefs of the chirurgic art." *Obfervations on the mortifications of the toes and feet, page* 793.

The Gout in the Foot or Hand.

No. 460. *Apply a raw lean beef-fteak. Change it once in twelve hours till cured.*

Inftead of making any remarks of my own upon this curious remedy, I fhall only here take the liberty of tranfcribing what hath been faid in relation to it by the Rev. Mr. TOPLADY. " In Mr. Wefley's " book of receipts, entitled *Primitive Phyfic,* he ad- " vifes perfons who have the gout in their feet or " hands, to apply raw lean beef fteaks to the part " affected, frefh and frefh every twelve hours. Some- " body recommended this dangerous repellent to " Dr. T. in the year 1764 or early in 1765. He " tried the experiment; the gout was, in confequence, " driven up to his ftomach and head, and he died

" a

" a few days after *at Bath*, where I happened to
" spend a confiderable part of thofe years ; and
" where at the very time of the Dean's death, I
" became acquainted with the particulars of that
" cataftrophe.

" I am far from meaning to infinuate, becaufe I
" do not know, that the perfon who perfuaded Dr.
" T. to this fatal recourfe derived the recipe imme-
" diately from Mr. Wefley's medical compilation.
" All I aver is, that the recipe itfelf is to be found
" there, which demonftrates the unfkilful temerity,
" wherewith the compiler fets himfelf up as a phy-
" fician of the body. Should his quack pamphlet
" come to another edition, 'tis to be hoped that the
" *beef fteak* remedy will, after fo authentic and fo
" melancholy a probatum eft, be expunged from
" the lift of fpecifics for the gout.—'Tis, I acknow-
" ledge, an effectual cure. Cut off a man's head, and
" he'll no more be annoyed by the tooth-ach ;
" Alas, for the *ingenium velox*, and for the *audacia*
" *perdita*, with which a rafh empiric, like Juvenal's
" *Græculus efuriens*, lays claim to univerfal fcience !

" *Grammaticus, Rhetor, Geometres, Pictor,*
. *Aliptes.*"
" *Augur, Schænobates,* Medicus, Magus" om-
nia novi.t *

Mr. Toplady alfo obferves, with reference to Mr.
Wefley, " *Aliquis in omnibus, nullus in fingulis.* The
" man who concerns himfelf in every thing, bids
" fair not to make a figure in any thing. Mr. John
" Wefley is, precifely this, *Aliquis in omnibus.* For
" is there a fingle fubject, in which he has not en-
" deavoured to fhine ?—He is alfo, as precifely, a
" *Nullus in fingulis.* For has he fhone in any one
" fubject which he ever attempted to handle ?†"

* Preface to the Scheme of Chriftian and Philofophical Ne-
ceffity, p. 9.

† Scheme, p. 9.

No.

No. 478 to 501, Mr. W. gives a variety of external and internal remedies for different kinds of the head-ach, upon which I shall only observe, that as pains attacking any part of the head must take their rise from some cause, so it is impossible that any person can prescribe judiciously, without considering whether it be a febrile, an inflammatory, or a rheumatic pain in the head; or an affection of the brain or nervous system; or whether the head-ach arises from a disorder of the stomach or bowels. These circumstances should be duly considered, before any one can, with the least degree of propriety, attempt to prescribe remedies for the various disorders of the head.

Heart-Burning.

No. 502 to 509, Mr. W. has prescribed a number of trifling recipes for the heart-burn. This complaint generally takes its rise from a weak and relaxed state of the digestive organs, and therefore it would have been better if Mr. W. (supposing him to have had any knowledge of the subject) had prescribed Tonics, which by strengthening the stomach, bid the fairest for removing this troublesome complaint.

Hoarseness.

No. 515, *Rub the soles of the feet before the fire, with garlick and lard, well beaten together, over night. The hoarseness will be gone the next day.*

This is a very extraordinary prescription, but as such a very extraordinary character is given of its certainty of success; it is to be hoped that every person affected with a hoarseness, will rub the *soles of his feet*, " with garlick and lard" as by so doing, a disorder of the *throat* " will be gone the next " day."

Mr. W. has given such a farrago of absurd remedies for the various diseases for which he pretends

to

to prescribe, as are enough to exhaust the patience of any ordinary reader; but my duty to the public obliges me to proceed, notwithstanding the irksomeness of the task. To those afflicted with pains in the joints, he advises, (No. 539) that they should *drink a decoction of herb-robert, and apply it as a poultice.* Now, pains in the joints may arise from causes very different, and yet the pain which is only a symptom of the other diseases, is to be cured by a single herb. But in order to render herb-robert the more certainly efficacious, it is to be applied both internally and externally. The egregious quackery of all this is too manifest to need any further remarks.

The Itch.

No. 540 to 549. If there be any disorder which Mr. Wesley understands, it appears to be *the Itch*; whether this be the result of his own feelings or experience, or of any other cause, I pretend not to determine; but his remedies for this cutaneous disease are more judicious than almost any other in his book.

The King's-Evil.

From 550 to 558. Here are eight remedies for this inveterate disorder; but they all appear superfluous; for at No. 426 he has recommended a mercurial preparation, which he says will cure the King's-Evil in *forty days.* This is a very expeditious remedy for so stubborn a disease, so that if any dependance were to be placed on what Mr. W. first recommended, there would be little occasion to have recourse to any other prescription.

The

The Legs inflamed.

No. 560. *Apply fuller's-earth spread on brown paper. It seldom fails.*

No. 561. *Or bruised turnips.*

No. 562. *Or boiled turnips mixed with mutton fat.*

No. 563. *Or rub them with warm juice of Plantane.*

These are external applications for *inflamed legs*; but persons so afflicted, should take care how they repel such appearances as external inflammation ; for in acute diseases it is often a very happy termination or crisis of a very long and dangerous fever, &c. and in chronic disorders it will be at all times adviseable to mend the habit of body, before an attempt be made to remove this inflammatory symptom.

Legs sore, and running.

No. 564. *Wash them in brandy, and apply elder leaves, changing them twice a day.—This will dry up all the sores, though the legs were like honey-combs. Tried.*

No. 565. *Poultice them with rotten apples.* *Tried.*

If it were to be admitted that these *tried* remedies had all the efficacy in them that Mr. W. attributes to them, yet they should not be used but with great caution. For it is an established law in the human body, that when any discharge, from whatever cause, has continued any length of time, it then becomes *habitual to the constitution ;* and therefore any astringent or repellent, which will cause a sudden check of the humors, will often be attended with the worst consequences to the general health; so that it is hardly ever safe to dry up *suddenly* a considerable flux of humours determined to any part, unless the habit

be

be mended by an alterative courſe, or an artificial out-let be made by an iſſue or ſeton, to unload the conſtitution, upon the drying up or cure of ſuch running ſores.

The Lethargy.

From 575 to 578. Mr. Weſley gives ſeveral preſcriptions for the cure of the lethargy, and among the reſt he orders *white hellebore to be ſnuft up the noſe.* This may prove a very dangerous remedy, as the lethargy is generally owing to an over-fulneſs of the blood-veſſels, and particularly thoſe of the head : now, any errhine, and eſpecially one ſo powerful as hellebore, might, by its ſudden and violent ſtimulus, cauſe an inſtant rupture of the veſſels of the brain, and an apoplexy be the unhappy conſequence. The lethargy is a complaint owing to an internal cauſe, and ought to be very ſeriouſly attended to, as it is often the forerunner of diſeaſes of the moſt dangerous nature, ſuch as apoplexy, palſy, &c.

No. 584 to 600. As Mr. W. is a univerſal practitioner, he preſcribes for *lunacy, raging madneſs,* and the bite *of a mad dog,* as well as for other diſeaſes ; but unleſs the friends and relations of the unhappy perſons ſo afflicted, are as mad as the patients, they will apply for proper advice and aſſiſtance, inſtead of relying on the modes of cure recommended by the author of Primitive Phyſic.

The Meaſles.

No. 601. *Drink only thin water-gruel, or milk and water, the more the better ; or toaſt and water.*

No. 602. *If the cough be very troubleſome, take frequently a ſpoonful of barley-water, ſweetened with oil of ſweet almonds, newly drawn, mixt with ſyrup of maidenhair.*

The

The prescriptions are only suitable drinks, with a little oil and syrup to palliate the cough; but Mr. W. does not inform his readers that this infectious disease is always attended with considerable inflammation of the breast, and that the lancet is frequently to be used to remove the inflammatory affection of the lungs, as well as to prevent the future bad consequences of the measles. Nor does our author recommend any kind of physic to be given at the going off of the disease; though clearing the constitution of the remaining morbillous matter, is a circumstance of the utmost importance with regard to the general health. This disease, if it be attended to by a judicious practitioner, never turns out ill, either during its continuance or afterwards : but if proper precautions are not used during the inflammatory state of the disease, it generally settles upon the lungs ; and thus, through neglect and inattention, vast numbers of children are carried off by the measles every year. No person ought then to slight the cough or other remaining symptoms of this disorder, although the patient should have been freed from the disease for a considerable time; as these are the warnings of the impending danger.

From 604 to 612, are a great variety of remedies advised for *menses obstructed*, but no attention is paid to the age or the constitution ; so that it is to be hoped, that the fair sex will pay little attention in this case to the recipes contained in the *Primitive Physic*; as it is not to be supposed that Mr. W.'s *female auditory* have consulted him much in this complaint. And if it be injudiciously prescribed for, it may lay the foundation for a future ill state of health, and for disorders which never can be removed.

Mr. W. has also from 612 to 620 sundry prescriptions for *menses nimii*. I confess myself somewhat at a loss to know what were the reasons for using the word *nimii*. The word *menses* is, I believe, generally

nerally underſtood; but how the good women in the country are to comprehend the other word, I know not. They may indeed apply to the parſon of the pariſh; but ſhould their delicacy prevent this, or the parſon not happen to be in the way, they may flow on for the next month, before the female patient may know what theſe excellent recipes are good for. However the complaint referred to, ought not to be trifled with, but a due regard ought to be paid to age, conſtitution and other circumſtances, to prevent the bad conſequences reſulting from the diſorder.

Old Age.

No. 629. *Take tar-water morning and evening. Tried.*

No. 630. *Or, decoction of nettles; either of theſe will probably renew the ſtrength for ſome years.*

No. 630. *Or, be electrified daily.*

Mr. Weſley, who is a moſt incomparable practitioner, has remedies for a diſeaſe, of all others the moſt inveterate, viz. *old age.* *Tar-water* is a *tried* remedy; or if that *tried* preſcription ſhould be found not ſufficiently efficacious, *decoction of nettles*; and " either of theſe," he ſays, " will probably renew the ſtrength for ſome years." Or if the patient ſtill feels old age an inconquerable diſorder, he recommends being *electrified daily.* This hint is worthy the attention of the ingenious Dr. Priestly; as when the arcana of electricity are compleatly laid open, an electrical ſhock judiciouſly adminiſtered, and repeated with ſufficient frequency, might peradventure extend a man's life to a thouſand years; or if it were only *five hundred*, it might be as advantageous to the public as Dr. Priestly's diſcoveries reſpecting fixed air; though theſe have juſtly intitled
this

[63]

this gentleman to that applaufe which he hath univerfally received in the philofophic world.

No. 652 to 664. Thefe are external applications
for the cure of the piles; but they deferve little regard. This is one of thofe diforders, which though
very painful, generally tend to do great good to the
conftitution ; for when a perfon is of a fanguineous
or melancholic temperament, or his veffels act
very ftrongly, this is the moft happy determination
that nature can take ; and although the piles may return feveral times, and no evacuation of blood be
occafioned, yet the ftimulus is tending that way ;
and if the pain and uneafinefs fhould be removed by
repellents or aftringents, it will often be at the hazard
of the life of the patient. I have been a melancholy
eye-witnefs of the truth of this affertion, in two or
three cafes in my own practice, where perfons, from
the uneafinefs they fuffered, have (contrary to my advice) rafhly applied fome aftringent, which has fuddenly removed the piles, and made a very flight and
falutary complaint terminate in an apoplexy.

The Pleurify.

No. 675. *Apply to the fide onions roafted in embers
mixed with cream.*

No. 676. *Take half a dram of foot.*

No. 677. *Take out the core of an apple, fill it with
white frankincenfe : ftop it clofe with the piece you cut
out, and roaft it in afhes. Mafh and eat it.*

No. 678. *A glafs of tar-water, warm, every half
hour.* -

No. 679. *Decoction of nettles ; and apply the boiled
herb hot as a poultice. I never knew it fail.*

No. 680. *Boiled fennel, or camomile-flowers.*

Soot

Soot and tar-water are the only internal remedies advised by Mr. W. for the cure of the pleurify, except his *frankincenfed apple* ; he has indeed feveral prefcriptions as external applications to the fide. In almoft every fection the author of the Primitive Phyfic proves to a demonftration his ignorance of the animal œconomy, and even of the firft principles of medical knowledge ; but in no one more than the prefent ; as he undoubtedly means that his recipes fhould be employed for the cure of the inflammatory pleurify. Mr. W. appears to have hardly any idea of any thing in medicine, but removing pain, not confidering that the pain in the fide, in this cafe, is one of the fymptoms of an inflammation of the pleura ; and if the lancet is not freely ufed the firft two or three days of the difeafe, it will terminate in an internal abfcefs, which formation of matter in the breaft will moft commonly kill the patient. Mr. W. in a note defines a pleurify to be *a fever attended with a violent pain in the fide, and a pulfe remakarbly hard.* But is it not extraordinary that he fhould give fuch a definition of the difeafe, and not order bleeding, which is a certain method of removing the hard pulfe, as well as the other fymptoms of inflammation ? But inftead of advifing evacuation in an inflammatory difeafe, the author of the Primitive Phyfic prefcribes *foot,* tar-water, and *frankincenfe.*

To one poifoned.

No. 683. *Give one or two drachms of diftilled verdigris, it vomits in an inftant.*

Mr. Wefley directs, that to one poifoned, fhould be given ONE or TWO *drachms of diftilled verdigris.* This deftructive prefcription was juftly animadverted upon by a fenfible writer, who figned himfelf ANTIDOTE, in the Gazetteer of Dec. 25, 1775, and it was
this

this gentleman's obfervations which firft led me to perufe Mr. W.'s *Primitive Phyfic* ; and which ac- cordingly gave rife to thefe remarks upon that pub- lication. In the letter referred to, Antidote fays, " Every one who has the leaft acquaintance with the " powers of medicine, will, I believe, be equally " ftartled with myfelf at reading fuch a prefcription. " I could fcarce believe my eye-fight for fome time, " nor can at prefent by any means account for the " ignorance and prefumption of a man who deals " out as an antidote, one of the moft active poifons " in nature, in fuch an enormous dofe, and this " in fuch an undetermined quantity, as if the exact " dofe were immaterial." And Antidote further obferves, addreffing himfelf to Mr. W. " it is very " probable that your dofe of two drams would ef- " fectly poifon 20 or 30 people, or operate very fen- " fibly on every man, woman, and child, in one of " your largeft congregations."

Two drams of verdigris are indeed fufficient to poifon forty or fifty people, and that fuch a direction fhould have been given in a book intended for ge- neral ufe, and which has paffed through many edi- tions, is a moft alarming confideration, and ought to have given Mr. W. the greateft concern. But in anfwer to this charge he publifhed the following let- ter in the Gazetteer of January 1, 1776.

To the PRINTER of the GAZETTEER.

Dec. 28, 1775.

" Between twenty and thirty editions of the *Pri-* " *mitive Phyfic,* or, A Rational and eafy Method of " curing moft Difeafes, have been publifhed either " in England or Ireland. In one or more of thefe " editions ftand thefe words. " Give one or two. " drachms of verdigris." I thank the gentleman

I

" who

" who takes notice of this, though he might have
" done it in a more obliging manner.
" Could he poffibly have been ignorant (had he
" not been willingly fo) that this is a mere blunder
" of the printer? that I wrote *grains* not drachms?
" However, it is highly proper to advertife the
" public of this; and I beg every one that has the
" book, would take the trouble of altering that
" word with his pen.

" Your's, &c.

" J. WESLEY."

Mr. W. above fays, that this dangerous error ftands in *one or more of the twenty or thirty editions of the Primitive Phyfic, which have been publifhed either in England or Ireland.* But this appears to be a moft artful evafion; for this error is in the *fifth, the eighth,* and the *fixteenth* editions; and there is the greateft reafon to believe, that it has paffed through every edition; for though Mr. W. has been publicly called upon to point out the edition in which there was not this error, he has not been able to point out any one. He has indeed, with a jefuitifm truly characteriftic of himfelf, infinuated, though not afferted, in a letter inferted by him in the Gazetteer of Jan. 31, that this moft dangerous and fatal blunder was referred to in the *errata:* but this infinuation appears to be totally without ground, for I could never meet with fuch a correction in any edition, and if it had ever been difcovered before, it muft have been the moft culpable and fhamelefs negligence, to have fuffered fo fatal a prefcription to ftand in the laft edition.

In Mr. Wefley's firft letter, as given above, he afks, " Could he (*Antidote*) poffibly have been igno-
" rant, (had he not been willingly fo) that this is a
" mere blunder of the printer? That I wrote
grains,

grains, not *drams?*" This is, perhaps, the firſt
time that ever any author had the modeſty to cenſure
his opponent, for not taking it for granted that he
wrote right, when he was convicted of having writ-
ten wrong. But as *Fly-Flap*, another writer in the
Gazetteer, juſtly obſerves, " The weak attempt to
" throw the blame upon the Printer, is as uncandid
" as it appears improbable:" For, " the words,
" drams and grains, are ſo unlike, that it is almoſt
" impoſſible to miſtake the one for the other."

It might have been expected, that Mr. Weſley,
when he had diſcovered ſo dangerours a preſcripton
in his book, (a preſcripton which might be of ſuch
fatal tendency) would have been exceedingly alarmed ; and even, if he had been really deſtitute of the
feelings of humanity, that he would, however, have
pretended ſome concern, leſt the blunder ſhould
have been productive of ſome miſchief. But ſo far
from expreſſing grief, he appears to rejoice in the
hope, that the ſale of his pamphlet would be en-
creaſed by the attacks upon him, on this occaſion.

In his letter in the Gazetteer, Jan. 31, he ſays, " In
" one reſpect, I am much obliged to the Gentlemen,
" (or Gentleman) who ſpends ſo much time upon the
" *Primitive Phyſic* ; and would humbly intreat them
" to ſay ſomething about it, (no matter what) in half
" a dozen more of your papers. If nothing was
" ſaid about it, moſt people might be ignorant that
" there was any ſuch tract in the world. But their
" mentioning it, makes more enquire concerning
" it, and ſo diſperſes it *more and more.*"—Aſtoniſhing effrontery and inſenſibility !

If Mr. Weſley had conſidered the lives of his
fellow-creatures, as an object of much concern, the
leaſt he could have done, it might reaſonably be preſumed, would have been to have cancelled the leaf
wherein this dangerous blunder was, and to have
cauſed another to have been printed, and inſerted in

I 2   the

the unfold books. But he has done nothing like this ; he has only advertifed the error in one paper ; at leaft I have feen it in no more. Since he has been attacked on this fubject, I have caufed one to be bought in Paternofter-Row, which had in it this dangerous prefcription, not even altered with the pen. I have, indeed, fince fent for one to the Foundery, wherein the blunder was flightly corrected with the pen : but was this all that ought to have been done by Mr. W. as a man of humanity, or can his negligence, in this refpect, be judged confiftant with any due regard to the lives of his fellow-creatures ? Indeed, it is fomewhat extraordinary, that when the unexpected fuccefs of the *Primitive Phyfic,* had caufed Mr. Wefley, as he fays, *carefully to revife the whole, and to publifh it again, with alterations,* fo enormous a blunder fhould have paffed through all the editions ; for this appears to have been in fact the cafe. But the truth probably was, that Mr. W.'s ignorance firft occafioned this dangerous prefcripton, and the fame ignorance continuing, prevented it from being corrected in any of the editions. This however, fhews how little Mr. W.'s judgment is to be depended on ; and the little concern he expreffes for leading his readers into an error, which to fome may have proved fo fatal, is a ftrong evidence of his infenfibility. And when we confider the very extenfive fale of his book, the credulity of his followers, and the extreme ignorance which is manifefted in many of his prefcriptions, may we not fay nearly in his own words. *How many inconveniencies muft this have* occafioned! *How many conftitutions* may *hereby* have been *ruined! How many valuable lives have been loft!* *

The Quinfey.

No. 697. *Apply a large white bread toaft, half an inch thick, dipt in brandy to the crown of the head, till it dries.*

* Wefley's Preface, p. xxvii.

I am

I am fatisfied from experience, that exciting an inflammation upon the fkin, near the part affected, has done much good ; and even this prefcripton of *toaft and brandy*, might, perhaps, have been ufefully employed as a poultice to the outfide of the throat ; but if applied to the crown of the head, though it be repeated till doomfday, it cannot be of the fmalleft advantage.

A Quinfey of the Breaft.

702. *Take eight or ten drops of laudanum lying down in bed.*

The learned and ingenious Dr. Heberden, in the fecond volume of the Medical Tranfactions, among many other very valuable obfervations, treats of a new diforder of the breaft, which he calls *Angina Pectoris.* I fhall here take the liberty to quote fome of the Dr.'s judicious remarks upon this fubject, and the rather as the work, from which they are ex-tracted, is known to but few readers, except the faculty.

Page 59. " Thofe who are afflicted with it, are " feized, while they are walking, and more particu-" larly when they walk foon after eating, with a pain-" ful, and moft difagreeable fenfation in the breaft, " which feems as if it would take their life away, if " it were to continue or increafe : the moment they " ftand ftill, all this uneafinefs vanifhes. In all o-" ther refpects, the patients are at the beginning of " this diforder, perfectly well, and in particular, have " no fhortnefs of breath, from which it is totally " different."

" When a fit of this fort comes on by walking, " its duration is very fhort, as it goes off almoft im-" mediately upon ftopping. If it come on in the " night, it will laft an hour or two ; and I have met
" with

" with one, in whom it once continued for feveral
" days, during all which time the patient feemed to
" be in imminent danger of death.

" The pulfe is, at leaft fometimes, not difturbed
" by this pain, and confequently the heart is not
" affected by it; which I have had an opportunity
" of knowing by feeling the pulfe during the pa-
" roxyfm."

Page 66. " BLEEDING, vomits, and other eva-
" cuations, have not appeared to me to do any
" good. Wine or cordials taken at going to bed,
" will prevent or weaken the night fits; but nothing
" does this fo effectually as opiates. Ten, fifteen,
" or twenty drops of tinctura thebaica taken at
" lying down, will enable thofe to keep their beds
" till morning, who had been forced to rife, and
" fit up two or three hours every night, for many
" months. Such a quantity, or a greater, might
" fafely be continued as long as it is required: and
" this relief afforded by opium may be added to the
" arguments, which prove thefe fits to be of a con-
" vulfive kind."

Mr. Wefley, in many parts of his Primitive Phyfic,
proves himfelf an adept in plagiarifm ; and many
authors, there is no doubt, from whom he has
borrowed, would do him no credit, had he mentioned
their names ; but to have acted like a man of candour, he fhould have informed the public, that
the difcovery of this new diforder, as well as the mode
of treatment, was made by Dr. HEBERDEN, to whom
the honour of it ought certainly to have been attributed. I think I cannot conclude this fubject better
than by giving the Dr.'s own words relating to this
difeafe.—

Page 67, " Time and attention will undoubtedly
" difcover more helps againft this teizing and
" dangerous ailment; but it is not to be expected,
" that

" that much can have been done towards eftablifh-
" ing the method of cure for a diftemper hitherto fo
" unnoticed, that it has not yet, as far as I know,
" found a place, or a name in the hiftory of dif-
" eafes."

The Rheumatifm.

No. 703 to 715. Here are twelve remedies prefcribed for the cure of this diforder ; but moft of them
are fo extremely infignificant that they deferve no
attention, and the only one that feems likely to have
any action would in all probability prove highly injurious.

The rheumatifm has been commonly divided into
two difeafes, viz. the rheumatic fever and the chronic rheumatifm. The acute rheumatifm, or rheu-
metic fever, as it is commonly called, generally attacks young men, and thofe who are naturally of a
good conftitution. It is generally attended at the
beginning, with a hard, ftrong, full pulfe, and other
fymptoms of general inflammation in the habit ; at
the firft attack of this difeafe, copious and repeated bleeding can be the only ufeful remedy, which
if neglected for two or three days, or guiacum (in
fubftance, or the volatile tincture) or Mr. W.'s recipe, No. 708, *Steep feven cloves of garlick in half a
pint of white wine, drink it lying down* ; the heart, and
arterial fyftem would be ftimulated to fo great a degree, as to transfer the difeafe to the brain, or fome
other vital part, which often in a few hours proves fatal.
And the writer, with the greateft concern, declares,
that he is throughly convinced, many lives have been
loft by the common mode of prefcribing *guiacum*,
and other heating remedies, at the beginning of rheumatic complaints.

· No. 733 to 741. Are a variety of ftrange remedies advifed for the cure of the fciatica. One of
thofe curious prefcriptoms is, *a mud made of powdered pitcoal, and warm water.* But this mud will not
only cure the *fciatica,* but Mr. W. informs us, that it
alfo

alfo cures *palfies, weaknefs, weaknefs of the limbs, moft diforders of the legs, and fwellings and ftiffnefs of joints.* And alfo that it *cured a fwelling of the elbow joint,* though *accompanied with a fiftula, arifing from a caries of the bone.* How much is it to be regretted, that this *mud* is not better known, as it is fo admirable a cure for fo many incurable difeafes! Incredulous people, indeed, doubt the reality of thefe cures; but we have Mr. W's authority in their favour, and this will furely fatisfy all——except thofe obftinate people who require proof and evidence inftead of affertion.

A Sore Mouth.

No. 777. *Apply the white of an egg beat up with loaf fugar.*

No. 778. *Gargle with the juice of cinquefoil.*

No. 779. *Beat together a pound of treacle, three yolks of eggs, an ounce of bole armoniac, and a nutmeg of allum a quarter of an hour. Apply this to the fore part, or to an aching tooth. Tried.*

Mr. Wefley has here recommended feveral topical applications, but he did not confider, or probably was ignorant, that fores in the mouth, &c. frequently arife from internal caufes, and unlefs the conftitution is made better, fuch fores will feldom heal; or, if they fhould, the difeafe will frequently fix upon fome more internal part, and be much more dangerous than the original complaint.

The Strangury.

No. 840. *Ufe the cold bath.*

The utility of the cold bath in the ftrangury is by no means apparent, it being a diforder of the urinary paffages, which may arife from a great variety of caufes, that bathing feems very little adapted to cure.

A Sur.

A Surfeit.

No. 846. *Take about a nutmeg of the green tops of wormwood.*

A furfeit is a diforder which arifes from various caufes, and requires the attention of an able practitioner; but from whatever caufe it may proceed, it is not very likely to be removed by the *tops of green wormwood.*

To ftop profufe Sweating.

No. 847. *Drink largely of cold water.*

Drinking largely of cold water has by fome perfons been recommended, and particularly in fevers, to occafion fweating, and may fometimes have been advantageous; but Mr. W. it is apprehended, is the firft who ever advifed this remedy *to ftop profufe fweating.* But as he is a very uncommon practioner, he may be confidered as having the better right to prefcribe uncommon modes of treatment.

Swelled Legs.

No. 851. *Bathe them every morning in cold water, and take an eafy purge twice a week.*

No. 852. *Take wormwood, fouthernwood, and rue: ftamp them together, and fry them in honey till they grow dry: Then apply them as hot as you can bear.*

Mr. Wefley does not confider that *fwelled legs* are only one of the fymptoms of fome other difeafe, as the tumour may be a fymptom of inflammation, fever, rheumatifm, &c. Now in all thefe diforders the above prefcriptions may prove extremely injurious, and even in fome cafes deftructive to the patient.

A Swelled Throat.

No. 853. *Gargle with decoction of nettles.*

K

No.

No. 854. *Or of primrose leaves.*

As swelling in the throat is generally only a symptom of an inflammation, or some other affection attacking the throat, what Mr. W. has recommended can be productive of no good to the patient. But the author of *Primitive Physic* has given many proofs of the mischief that may accrue to the sick, from those who prescribe remedies for diseases, with the nature of which they are unacquainted. Thus in p. 115 of his work he has prescribed for the quinsey; p. 127 for a sore throat; and now he has a separate section for a swelled throat; though these are only symptoms of one and the same disease.

To fasten the Teeth.

To clean the Teeth.

To prevent the Tooth-ach.

To cure the Tooth-ach.

Mr. W. has prescribed *only* twenty-two remedies for the teeth; some of which are to *fasten the teeth,* some to *clean the teeth,* others to *prevent the tooth-ach;* and several very unaccountable remedies to *cure the tooth-ach.*

The writer cannot help embracing this opportunity, for the benefit of his readers, of giving an extract from the ingenious Mr. RUSPINI's *(surgeon-dentist, Pall Mall)* LITTLE TREATISE, just published, which affords the strongest proofs of his skill in his profession, as well as of his humanity and generosity to the public.

Mr. RUSPINI, among many other very sensible observations, interspersed through his pamphlet, says, that, " Persons of all ages should clean their teeth " constantly with proper dentrifices every night and " morn-

" morning, and never omit to wash their mouths
" well with water after eating ; otherwise particles of
" meat, sweetmeats, and fruits, or many other
" parts of our food, by remaining between or
" about the teeth, will lay a foundation for future
" mischief. By following this advice, the teeth of
" people in general, but especially of children, will
" be preserved from decay."

" If the teeth happen to be decayed or painful,
" it has been the general custom to send for the
" next tooth-drawer, who commonly has not any
" idea of cure, but by *extirpating a tooth* ; a practice
" replete with ignorance and barbarity, often fol-
" lowed by dangerous and sometimes fatal conse-
" quences, and therefore never should be perform-
" ed except in those cases where no other remedy
" promises any probability of success. As there are
" a great variety of causes productive of pain in the
" teeth and gums, so there are various means by
" which ease may be procured ; but these, like all
" other diseases to which the human frame is more
" or less liable, require the knowledge of skilful
" practitioners for their cure.

" Ladies of tender constitutions, during preg-
" nancy, are *often afflicted with pain in one or more*
" *teeth ; others* afflicted in a similar manner by
" colds ; others by nervous attacks. It would be
" as absurd to loose a tooth in hopes of ease, that
" should be painful from any of these, or many
" other causes, as for a gouty man to part with a
" painful toe, and suppose that his disorder would
" be cut off with it. Experience shews, that the
" causes and seats of pain are frequently in distant
" places : numbers of persons have been deprived
" of their teeth without receiving relief from their
" pain, and suffer a severe operation without a pos-
" sibility of benefit."

Extreme Thirst.

No. 890. *Drink spring water in which a little sal prunella is dissolved.*

As thirst is a symptom which is generally atten-dant on fevers, inflammations, rheumatism, dropsy, &c. it is only to be removed by the remedies proper for the cure of those disorders; but as this is a very troublesome and disagreeable sensation, the patient may be rendered much easier by sucking the *nitre, lozenges, tamarinds, lemon and sugar,* &c. but drinking gallons of water, or other fluids, will not remove this symptom when the fever is high, or any other acute disease is violent.

Torpor (or numbness) of the Limbs.

No. 887. *Use the cold bath with rubbing and sweating.*

As torpor or numbness is generally a symptom of an incipient paralytic affection, I am firmly of opinion, that the cold bath is exceedingly improper in this case, and may prove highly injurious. Indeed, it might be imagined, that Mr. W. himself was apprehensive of its bad tendency; for immediately after recommending the *cold bath,* he advises *rubbing* and *sweating.* Now as it is generally allowed, that internal stimulants are necessary as well as external applications to the numbed part, certainly then the cold bath as a remedy can be productive of no good effect.

The Tympany, or windy Dropsy.

No. 899. *Use the cold bath with purges intermixed.*

No. 900. *Mix the juice of leeks and of elder. Take two or three spoonfuls of this morning and evening. Tried.*

As

As the tympany is a difeafe refulting from a weak and relaxed ftate of the ftomach and bowels, and which occafions great quantities of air to be generated from the focd, &c. and collected in the firft paffages, fo as to diftend them to a very confiderable degree, there is not the leaft reafon to fuppofe that the cold bath or purgatives can be proper : but indeed it is fo difagreeable and troublefome a difeafe, that it will generally oblige the afflicted to apply for proper advice, and therefore there isthe lefs danger of mifchievous confequences refulting from Mr. Wefley's directions.

No. 923. *Drink tar-water morning and evening.*

No. 924. *A decoction of pimpernel.*

No. 925. *Take decoction of agrimony thrice a day.*

No. 926. *Or, decoction, powder, or fyrup of horfetail.*

An inward Ulcer.

As internal ulcers are the confequences of inflammation of fome vifcus, membrane, or mufcular part, which from neglect or improper treatment at the beginning of the diforder, frequently terminates in fuppuration, and requires the greateft medical fkill to prevent its being fatal to the patient ; as the ulcer when once formed, from a variety of internal caufes, is very apt to fpread, and the matter is thereby abforbed into the circulating mafs, fo as to ftimulate the heart and arterial fyftem, bring on an hectic fever, and prove fatal ; it is therefore extremely abfurd that Mr. W. fhould order fuch infignificant remedies to remove a difeafe, that is apt to be attended with fuch melancholy confequences.

To ftop Vomiting.

No. 943. *Apply a large onion flit to the pit of the ftomach. Tried.*

That

That a flit onion applied externally fhould be a good and a *tried* remedy for an internal difeafe, is fome-what extraordinary : but extraordinary remedies can excite no furprize, to any man who is well read in Mr. W.'s *Primitive Phyfic*, and who gives any de-gree of credit to the marvellous affertions which are contained therein.

No. 945. *Infufe an ounce of quickfilver in a large glafs full of water for twenty-four hours. Then drink the water : I.*

This is given as an infallible prefcription ; for the unerring letter *I* is affixed to it ; but as the writer is no friend to implicit faith, he is apt to entertain fome doubts of its efficacy. When a perfon is afflicted with a vomiting, according to Mr. W.'s rule, he muft wait 24 hours before he can apply his remedy ; and when it is procured, it is not very probable that any good effects can be produced from it, as not the ten thoufandth part of a grain will be diffolved in the menftruum, (the water) and will prove as infipid and inefficacious as the water would have done, be-fore the digeftion of the quickfilver and water had taken place.

Mr. Wefley prefcribes many remedies to ftop vo-miting ; but he never once recommends the mints, *camomile*, or carduus, though infufions of thefe fto-machic and bitter vegetables, are often very ferviceable in many complaints of the ftomach, fuch as ficknefs, retchings, and vomiting. Indeed, Mr. Wef-ley, in his wonderful performance, feems difpofed to fall out with the good old women, as well as with the apothecaries ; for the remedies that the former have and do often prefcribe with fuccefs, he has taken no notice of : but he has in fundry dieafes recommended things much more abfurd and inju-rious than what are advifed by them.

As

As we are on the fubject of vomiting, the reader will perhaps excufe me, if I make a little digreffion, relative to what has appeared to me to be a frequent caufe of a propenfity to vomiting. I have often obferved upon enquiry, when fent for to patients affected with complaints of the ftomach or bowels, that they had over-night, or the preceding day, *drank punch*; and from the effects, and the fmell of the matters thrown up, had every reafon to think, that inftead of lemon juice, the acid ufed was the vitriolic fpirit; and, indeed, it is hardly poffible for any one to know in fuch a farrago as punch, what kind of ingredients it confifts of.

I am firmly convinced, that there is hardly any thing tends fo much to weaken and relax the ftomach, and bring on various ill confequences, fuch as indigeftion, lofs of appetite, &c. as bad punch. And therefore I earneftly advife the *punch drinkers* to have their fruit, fpirits, &c. brought to them feparately, and to become *punch makers*; by which means they will know what the compound confifts of, and that it is not a compofition of bad fpirits, vitriol, &c. Indeed, I would recommend, that preferved *lemon juice* fhould never be ufed, as it is always running into new fermentations, and though not fo bad as vitriol, yet is by no means fo wholefome as the juice when fqueezed immediately from the fruit itfelf. As vomiting and other difagreeable fymptoms affecting the ftomach, may eventually injure the whole fyftem; becaufe the ftomach can never be long difordered, without mifchief arifing to the animal œconomy in general; therefore thefe hints may not be unworthy of attention, and if they are productive of any utility, there will be the lefs reafon to apologife for the digreffion.

Bloody

Bloody Urine.

No. 946. *Take a quarter of a pint of sheeps milk twice a day.*

Urine by Drops, with heat and pain.

No. 949. *Drink nothing but lemonade.* Tried.

No. 950. *Beat up the pulp of five or six roasted apples, with near a quart of water; take it lying down. It commonly cures before morning.*

Involuntary Urine.

No. 952. *Take a tea-spoonful of powdered agrimony morning and evening.*

No. 953. *A quarter of a pint of alum posset drink every night.*

Sharp Urine.

No. 955. *Take two spoonfuls of fresh juice of ground ivy.*

Suppression of Urine.

No. 957. *Drink largely of warm lemonade.* Tried.

No. 958. *Or a scruple of nitre every two hours.*

No. 960. *A spoonful of juice of radishes.*

No. 962. *Or of bruised mustard-seed.*

Mr. Wesley prescribes several remedies for *bloody urine, urine by drops, involuntary urine, sharp urine,* and *suppression of urine;* but these retensions and suppressions of the urine, &c. may be owing to some fault in the urinary passages, such as gravel lodging in some of these parts, the stone, the dropsy, the venereal disease, or even to inflammation or spasmodic affection of the kidneys, ureters, &c. so that the numerous prescriptions advised by Mr. W.
cannot

cannot poffibly be taken to advantage without knowing from what caufe it proceeds.

The Whites.

No. 977. *Live chaftly. Feed fparingly. Ufe exer-
cife conftantly. Sleep moderately, but never lying on
your back.*

No. 978. *Take eight grains of jalap every eight days.
This ufuall cures in five weeks.*

No. 981. *Make Venice turpentine, flour, and fine
fugar, equal quantities, into fmall pills. Take three or
four of thefe morning and evening.*

No. 983. *After a purging take about fifteen grains
of cerufe of antimony in white whine, twice or thrice
a day.*

Here are many recipes prefcribed for the *fluor-al-
bus.* This is a very troublefome difeafe, and as it
may be owing to a variety of caufes extremely difficult of cure; but Mr. W. has advifed feveral remedies, and not a few, which he feems to think *infallible,* for thofe diftempers which the faculty find the
moft ftubborn, and the moft difficult to remove.
And indeed, if it was as eafy for Mr. W. to perform
cures as it is to write recipes, he would be univerfally
allowed to be fuperior to a Huxham, a Mead, or a
Boerhave.

As the complaints mentioned from 1003 to 1012
are the object of furgery, and treat of various kinds
of *wounds,* for which it is impoffible to lay down any
general mode of treatment; as the dreffing muft be
varied according to the ftate of the wound, aud other
attendant circumftances, it cannot reafonably be expected that Mr. W.'s recipes would be of any ufe, and
it muft therefore be needlefs to comment on them.

L

Mr.

Mr. Wefley concludes his *Primitive Phyfic* with the wonderful cures performed by *cold-bath.ng, wafh-ing the head, water-drinking, electrifying,* and laftly *fafting-fpittle,* which, *outwardly applied,* he informs us, *fometimes cures blindnefs and deafnefs,* befides various other diforders; and, *taken inwardly, it relieves or cures cancers, the gout, the king's evil, the leprofy, the palfy, the rheumatifm, the ftone, &c. &c.* He feems indeed to have been rather profufe of his remedies, which is hardly confonant to his own fentiments. For in his preface he fays, "Experience "fhews that one thing will cure moft diforders, at "leaft as well as twenty put together. Then why "do you add the other nineteen?" Indeed, it feemed hardly neceffary that Mr. W. fhould publifh a book containing 1012 recipes, when, according to his account, the above *five* remedies will cure almoft every acute and chronic difeafe incident to the human body. Mr. W. fays, p. 154, that *cold bathing cures young children of convulfions, coughs, cutaneous inflammations of the ears, navel and mouth, vomiting, &c. &c.* But I am clearly of opinion, that if the cold bath be ufed in thefe various difeafes of young children, agreeable to Mr. W's recommendation, the lives of many children will be facrificed in confequence. In many of the complaints for which it is prefcribed, it is totally improper; and in others that are mentioned it can do no good. And even where it is proper, it requires a little more attention than Mr. W. pays to the fubject; and indeed even common nurfes appear to underftand this matter better than he does. For they are feldom or never fo abfurd as to dip a child that is afflicted with coughs, cutaneous or other inflammations, vomiting, &c. But to enter into a particular examination of every abfurdity advanced by Mr. Wefley, would be equally tedious to me and my readers. What has been advanced, it is prefumed, may be fufficient to fhew the futility of many

of

of his prefcriptions, the pernicious tendency of others, and his total incapacity to produce any medical treatife calculated to be of the leaft fervice to mankind.

A book that has paffed through fo many editions as the *Primitive Phyfic*, muft have been attended to by great numbers; and as the recipes in it are often fo injudicious, abfurd, and and fo ftrongly characterized by ignorance of the human body, and of the power and operation of medicines, they may have been productive of great mifchief. Thefe confiderations will, I hope, be confidered as a fufficient apology for this publication. I have no perfonal animofity againft Mr. Wefley, to whom I am totally unknown; nor have I been induced to engage in this performance, by any confideration, refpecting the part Mr. W. has taken in the political world. Every thing of this kind is foreign to the defign of this examination of the *Primitive Phyfic*. But I have ever wifhed to underftand the principles of the medical art, to be ufeful in my profeffion, and ferviceable to my fellow-creatures; and if this little piece be found by the candid and judicious to be of that tendency, I fhall not regret the pains I have taken. I am confcious of the uprightnefs of my intentions, and therefore hope to meet with indulgence from the public.

F I N I S.